When Birds Are Near

When Birds Are Near

Dispatches from Contemporary Writers

Edited by Susan Fox Rogers

Comstock Publishing Associates
an imprint of
Cornell University Press
Ithaca and London

First published 2020 by Cornell University Press

Printed in the United States of America

Library of Congress Cataloging-in-Publication Data
Names: Rogers, Susan Fox, editor.
Title: When birds are near : dispatches from contemporary writers / edited by Susan Fox Rogers.
Description: Ithaca [New York] : Comstock Publishing Associates, an imprint of Cornell University Press, 2020. | Includes bibliographical references.
Identifiers: LCCN 2020004197 (print) | LCCN 2020004198 (ebook) | ISBN 9781501750915 (paperback) | ISBN 9781501750939 (pdf) | ISBN 9781501750922 (epub)
Subjects: LCSH: Birds—Literary collections. | Bird watching—Literary collections. | Essays. | LCGFT: Essays.
Classification: LCC PN6071.B55 W48 2020 (print) | LCC PN6071.B55 (ebook) | DDC 808.8/3628—dc23
LC record available at https://lccn.loc.gov/2020004197
LC ebook record available at https://lccn.loc.gov/2020004198

For Alice and Thomas, readers both. I'll make you birders yet.

Contents

Acknowledgments

I want to first thank my writers, some of whom stuck with me through a long editing process—I am grateful to all of you for your time and generosity in being a part of this collection. Your writing has inspired me and often made me dream of birds and far-flung destinations. I now can't imagine anything more perfect than birding with all of you.

My energetic and insightful editor, Kitty Liu, has been enthusiastic from the start and inspiring and so often funny through all of the (at times boring) details of putting together this collection. All of the wonderful people at Cornell, including Alexis Siemon, Susan Specter, and the copyeditor, Marian Rogers, have made editing this book not just possible but a pleasure.

My colleagues and friends at Bard College are too many to thank, but for all who have sent me a bird poem, question, observation, great essay, or musing, you have no idea how that has cheered my days. Dinaw Mengestu, Phil Pardi, Ellen Driscoll, Wyatt Mason, and Corinna Cape helped in magical ways to make this book happen. To my many birding friends—may I keep learning with and from you. Deb Addis read and coached from afar and Teri Condon from near.

My sister, Rebecca Rogers, listened to me talk about this book, read drafts, and now brings her binoculars on all our walks; she is the rock that makes all things possible.

A Note on Common Names

The essays in this collection follow the American Ornithological Society's system for capitalization of common names of birds. For example, Hermit Thrush instead of hermit thrush; Snowy Owl instead of snowy owl. When not using the specific name, however, the term is lowercased (e.g., the thrush; an owl). Capitalization style used in previously published essays has been retained.

When Birds Are Near

Introduction

Susan Fox Rogers

Some pig. Terrific. Radiant. Humble.

Charlotte, the marvelous spider of E. B. White's *Charlotte's Web*, weaves these words into her web in order to save Wilbur the pig from being carted off to the pig factory. I wasn't much of a reader as a child unless the books contained outdoor exploits—hikes or rafting trips, explorations of unknown lands (Enid Blyton was a favorite)—or if they focused on animals. So *Charlotte's Web* was a much-read book, and White a much-loved author. White did not write his wonderful book to teach a lesson. Still I walked away knowing this: a *spider* can save a pig's life. All it takes is imagination and some web-weaving skills. What might I do?

As an adult I keep E. B. White near, and I still turn to his essays when I need a cleansing, a reminder of how beautiful a sentence can be when you stick with nouns and verbs. And so it happened that one summer, not long into my bird-obsessed life, I sat down with White's essay titled "Mr. Forbush's Friends," and a new bird world opened to me. The essay introduced me to Edward Howe Forbush who is best known for writing *Birds of Massachusetts and Other New England States*, a three-volume set published in 1928. I think of guidebooks as necessary but not necessarily entertaining reading, so White's enthusiasm for Forbush's work intrigued me. E. B. White picks up one of Forbush's volumes when he is "out of joint, from bad weather or a poor run of thoughts." White makes Forbush seem like a bird explorer, a daring enthusiast and a real nut. The kind I want to be my friend. Forbush was often in precarious situations. like when his boat ran aground off Cape Cod, a sou'wester blowing, and his oars "carried off to sea." Does he panic? No, he is "absorbed in 'an immense concourse of birds' resting on the sands, most of them common terns." Forbush's passion for birds was singular and complete. White admired that "birds being near, Mr. Forbush found the purest delight."[1]

I was able to download Forbush's work, but my pleasure in reading him came only when a friend sent me the three-volume set for my birthday. The hardbound books are dull green and oversized, a pleasant weight on my lap while I read. Within these pages, I found the expected information—breeding and feeding, size and color—and a bit of the unexpected in his reports on the "Economic Status" of each bird. In this section, he offers how the birds are perceived in the human economy, like the Black-crowned Night Heron, which "is accused of being injurious to the fishery interests." Or Gannets, who "have been accused of doing considerable damage to the fishing interests." Forbush always defends the birds: "These harmful effects [of overfishing] have been much overestimated." And "There has been no thorough investigation of their food habits," he writes of the herons. Often, of course, no complaints can be made against a bird, as with the bitterns. Of them, Forbush writes: "Their economic status is not well known, but doubtless they are indispensable aids in keeping true the balance of life." Which seems another way of saying: Leave them be, let nature take its course.[2]

But what White enjoyed about Forbush was the section titled "Haunts and Habits." There, Forbush details his encounters with the species, or those of others who write in with their reports. Among the "Haunts and Habits," it's hard to beat his four-page description of spending a night in a Black-crowned Night Heron rookery near Barnstable. "On the ground under these trees," he noted, "the odor of ancient fish and that of the ammoniacal fumes accompanying decay were so nauseating that, having taken a few hurried snap-shots, I was ready to seek the open air to alleviate certain disagreeable symptoms."[3] Yet near this miasmic smell, he settled in for the night, head-net protecting him from the flies, midges, and mosquitoes that had swarmed around him all day long, ignoring a few hungry wood ticks that were still burrowing into his flesh. Most people would at this point flee, but not Forbush, who spends the night listening to the pandemonium of the rookery. Forbush manages to ignore, perhaps even enjoy, danger and discomfort as he brings himself closer to the birds.

Forbush was an inspiration in wanting to edit this collection, which I think of as an extended "Haunts and Habits" of contemporary birders. Hearing of others' encounters with birds always cheers me—nothing makes me happier than a friend or a student pulling out his or her phone and showing me a photo or video and asking, "What is this?" or "What is this bird doing?" These reports from the field create a collage of information that slowly adds up to some understanding. I wanted more of these stories from the field—in-depth reports that revealed the birds, our relationship to them, and perhaps also some unexpected wisdom about the mess of life, or this beautiful planet we live on.

To compile this work, I read widely. I relished Florence Merriam Bailey in the early twentieth century birding on horseback in the West, and was mesmerized by Kenn Kaufman hitchhiking his way through a Big Year in 1973. Narratives, old and contemporary, have given me new ways to think about birds, our relationship to them, and the ways the (bird) world has changed, both good and bad. I liked that we had moved from shooting birds to identify them to using an opera glass to focusing high powered binoculars. I didn't like that the thousands of Passenger Pigeons described by Alexander Wilson are now extinct. I liked that we no longer used bird feathers for women's hats; I didn't like that we have destroyed most of our native grassland and with that seen populations of grassland birds plummet.

Then I sent out calls for stories. Soon, dispatches arrived from the field describing bird life and behavior from Florida to Alaska, and beyond, and about birds that range from the Baird's Sparrow to the Sandhill Crane to the Great Skua. I delighted in those writers who seemed to adopt a

Forbushian attitude, relishing or dismissing physical challenges as the thrill of bird finding takes over: journeying by bus and foot along Highway 1 to find California Condors, or jouncing students down dirt roads to find one special bird, a Lucifer's Hummingbird. A masochistic approach to birding was not a prerequisite for inclusion in this collection, however.

What I was looking for in these essays I distilled to "birds plus something." These could not be simply guidebook reports from the field; the person writing had to add his or her unique perspective on the bird. Often that "something" is our role in the natural world, from the small to the large, from saving an injured bird to fighting the destruction of our grasslands. Sometimes that "something" is less concrete, more personal. As Rob Nixon writes in his essay, "Spotted Owls," "Sometimes the best bird experiences are defined less by a rare sighting than by a quality of presence, some sense of overall occasion that sets in motion memories of a particular landscape, a particular light, a particular choral effect, a particular hiking partner." Or, seen through Elizabeth Bradfield's lens in "Buried Birds": "We resonate with certain animals, I believe, because they are a physical embodiment of an answer we are seeking. A sense of ourselves in the world that is nearly inexpressible."

These essays are, then, not just field reports. They expand with reflections on love, family, life, and death and engage a range of emotions from wonder to humor. And because birds magnify our relationship to the natural world, you will read stories about habitat loss, declining species, birds that collide with buildings, or birds now extinct. Some too tell of small victories.

This wide-ranging collection weaves tales that show us *some bird. Terrific. Radiant. Humble.* It's a perfect read for a winter night when the wind is blowing and you are feeling out of sorts; it's an anthology to keep near when the birds are not.

Chapter 1

Nighthawks

Lake Perez

Katie Fallon

The sounds of a party—a wedding—float across Lake Perez in Stone Valley. Bursts of laughter, children calling to each other, and the low, bubbling murmur of conversation. I don't know who's gotten married, or where they're from, but love is in the air. And this is my favorite time of year: mid May. Appropriate for weddings, for renewal, for rushing into bloom.

I sit on the opposite bank, just down the hill from the cabin where I'll be spending the night. Before me, nighthawks wheel and boomerang above the lake's glassy black surface, their long wings cutting the air. The birds' movements are fluid, elastic, easy, and graceful. They swoop low, then climb, swoop low again, like giant, agile bats, hawking insects. Each long wing bears a distinct white stripe, which looks like a strip of reflective tape from below, and each bird's white throat patch gleams against the darkening sky. Perhaps the nighthawks are fueling up before the storm we all know will come tomorrow, the rain so common in an Appalachian spring. Perhaps they're pushing on ahead of it, migrating still, making their way north from their South American wintering grounds. I want to call across the lake to the folks at the wedding, to get them to look at the bird show overhead. Instead, I watch in wide-eyed silence.

Nighthawks and their relatives—whip-poor-wills, oilbirds, frogmouths, pauraques, and nightjars—are odd, secretive, mostly crepuscular or nocturnal birds. On the wing, a Common Nighthawk is acrobatic and incredibly sleek. In the hand, however, its wings seem too long, its body squat and strange, its eyes dark and clear as a mountain lake at dusk. A nighthawk's tiny black beak hides an enormous mouth that resembles a bullfrog's when it opens. Because they eat and drink while flying, this oversized mouth is useful for trapping insects and skimming lake water. The ancient Greeks and Romans believed that these unusual birds used their huge mouths for another purpose: drinking milk from the teats of goats and sheep under the cover of night. According to the lore, a goat suckled by a nightjar met an unfortunate end—blindness and then death. Of course, the birds do not engage in this behavior, but the belief earned their family the name Caprimulgidae, or "goatsucker."

As Stone Valley darkens, I retreat to my cabin and recline on the bench outside the door. Birds around me sing to the fading day: an Eastern Wood-pewee (the first I've heard this spring), a Wood Thrush, and Chipping Sparrows below the pines. Frogs along the lakeshore join the chorus, but my mind is still soaring with the nighthawks. My first encounter with a nighthawk had been more than fifteen years earlier. I'd just started graduate school and had moved to West Virginia with my boyfriend (now my husband), Jesse. He dreamt of going to veterinary school one day, so two evenings a week he volunteered at a local small animal clinic. We also began volunteering together at a wildlife rehabilitation center, and injured birds of all sorts began to find their way to us. Shoeboxes and dog carriers would appear at the clinic, containing limping geese, twisted ducklings, cat-attacked robins, and, one evening, a small bundle of brown and black feathers with long wings, a mini beak, and glossy black eyes. Someone had found the strange bird stunned on the shoulder of a road and scooped it into a box. Radiographs showed a wing fracture, but it wasn't badly displaced. We wrapped the wing to the bird's body and would wait for it to heal.

We soon realized that caring for an immobile nighthawk would be difficult. Three or four times a day I cupped the bird in my hands while Jesse gently pried open its beak and pushed a cricket or mealworm or soggy piece of cat kibble inside. It was labor intensive and stressful for the bird (and us), but we all soldiered on. I remember how warm the bird was, how its feathers were impeccable. Jesse and I worried it would lose too much weight, or that our insect-and-cat-kibble regime wasn't appropriate. We kept the box clean and warm, lined with soft cloth. We cooed over the bird, we stared into its black eyes. Of course, we fell in love. With the nighthawk, and with each other.

Weeks passed. Finally the bone was stable, calloused, and it was time for the bird to exercise. But how? The wildlife center didn't have a flight cage with small enough mesh, and the veterinary clinic didn't have a spare room. Our apartment was too crowded with animals already. So we improvised. Behind the animal hospital was a wet, swampy meadow, filled with high grass and cattails. At dusk Jesse and I would head out there, stand facing each other, and slowly, gently, toss the nighthawk back and forth. Every evening we stood further and further apart, and the bird's strength returned. The last few evenings, it wheeled over our heads, and we turned and sprinted after, following the bird to the place it finally landed. Then one evening it happened; I gently tossed the nighthawk, and the bird beat its long wings and lifted, lifted, lifted into the darkening sky, much higher than it had flown before. Jesse ran but it was futile. The nighthawk kept going, higher and farther, until it was out of sight. We cheered and cried, hugged, and collapsed, laughing in the meadow.

From my bench outside my cabin in Stone Valley, I smile at the memory and look out over Lake Perez. Fish lip the water, leaving concentric rings on the surface. The robins settling in the pines sing abbreviated songs. The wedding's voices and laughter continue to float across the lake, though muted now, softening. Nighthawks still dance in the twilight, their reflections flickering on the dark water. I will never know for sure if our nighthawk's repaired wing was strong enough to fly to South America and back, season after season. Perhaps the bird ended up on a road again, or succumbed to any one of a number of dangers during migration. Perhaps, ultimately, the life of one nighthawk is insignificant; perhaps our human lives are insignificant, too. But no matter how small, on that day's end, as the sun slipped below the horizon, what returned to the sky was made of love, was buoyed by love. The same love spins in the air tonight and fills the valley. Long may it fly.

Chapter 2

Spotted Owls

Rob Nixon

October 3, 1995: verdict day in the O.J. Simpson trial. A day that every American of a certain age recalls—where they were, with whom—when law clerk Deirdre Robertson pronounced the words "Not guilty" at 10:07 a.m. Pacific time. I too remember my exact location: up Scheelite Canyon in the Huachucas, looking for Mexican Spotted Owls, with Robert T. Smith for company.

The Spotted Owl flew across my sightlines as a political football long before it had any reality as a living bird. Like most people living in the U.S. during the late 1980s and early 1990s I had never heard a Spotted Owl's high-pitched four-note bark, but I had heard, ad nauseam, politicians cawing and squawking about this creature they reviled or revered.

The one-and-a-half pound owl became an inadvertent celebrity. It took flight into the symbolic stratosphere, becoming more or less the polar bear of America's temperate old-growth forests. The Spotted Owl emerged as an indicator species not just of forest health, but of a fevered nation's political temperature. This mild-mannered, elusive bird incited all kinds of fractiousness: arguments about jobs vs. the environment, about clear-cutting vs. sustainable logging, about absentee owners vs. local control of forest management, about what exactly "endangered" means. Café proprietors in irate mill towns added "spotted owl soup" to their menus. Bumper stickers exhorted Americans to "Save a Logger, Eat an Owl." The bird's fate provoked legal fisticuffs between two federal agencies, the Bureau of Land Management and the Fish and Wildlife Service. By the early nineties the Spotted Owl seemed to have migrated opportunistically from the ancient forests it had favored historically to a whole new ecological niche in the federal court system.

But when I traveled to southeast Arizona in early October 1995, owls were far from my mind. What drew me to that corner of the state were birds at the opposite ends of the ornithological spectrum: ostriches and hummingbirds. At the time, I was researching a memoir about ostriches. I'd grown up near a South African town that in the beginning of the twentieth century had boasted 10,000 people and close to a million domesticated ostriches, courtesy of the Edwardian obsession with voluminous, feathery fashion. By the mid-1990s, Wilcox, Arizona, had become an epicenter of American efforts to reprise that bygone era of money-spinning ostriches. But after a few weeks spent in the company of Arizona's big bird boosters I'd reached ostrich saturation point and, for a species break, was venturing to Ramsey Canyon to do a travel piece on the area's prolific hummingbirds.

So in southern Arizona I had my sights set on the flightless and the superflighted, ground-bound behemoths and minuscule birds hovering at 300 wingbeats a minute. Spotted Owls weren't part of the equation.

On my second day at the Ramsey Canyon Inn, I started asking around for an expert on the area's birds. The response was uniform: "You should speak to Smitty." So I got his number, called, and asked if I could do an interview, not knowing what to expect.

"Interview? I'd rather take a hike. Anyway, I only have one chair at home so there'd be nowhere for you to sit. Tell you what, why don't I take you up Scheelite Canyon to look for Mexican Spotted Owls? Sighting pretty much guaranteed."

He might as well have offered to lead me to a breeding pair of unicorns with a foal in tow.

At precisely 6:00 a.m. the following day, I met Robert T. Smith near the military checkpoint for Fort Huachuca through which we had to pass to gain access to the canyon. Smitty—a sinewy man in his seventies—didn't talk much at first, but was emphatic when he spoke. He was glad to get away from the hoopla of the O.J. trial. So was I, for my own reasons. The LAPD's rank racism, the lies on all sides, the cloying theater of it all. But my reasons for going AWOL ran deeper than that. I have a dead spot in my brain where a proper appreciation of celebrity ought to be. In my case, something about that essential faculty has never been fully operational. I put it down to growing up in a culturally isolated South Africa where the apartheid regime had outlawed all TV. Celebrity was an exotic notion that I first encountered in America in my twenties, far too late for it to achieve any reliable emotional traction. Despite my disciplined efforts at cultural adaptation, celebrity attentiveness remains for me a challenging exercise. Birds—which I encountered at a far younger age—draw me in much more easily.

Smitty warmed to his subject as we walked, as if his legs brought some essential lubricant to the conversation. It was clear the owls excited him. He talked with quiet pride about his achievements, which included banning the playing of Spotted Owl recordings on all U.S. military properties. Ex-military himself, he expressed contempt for tape-playing birders who put their lust for a lifer above the owls' right to remain undisturbed. Smitty had become the trail's unofficial custodian, dragooning Fort Huachuca staff into maintaining the path that led us up a narrow, deeply shaded canyon.

As we approached the owls' favored haunts, Smitty talked, sotto voce now, about their vulnerability. To deforestation up north. To predation by Red-tailed Hawks, Northern Goshawks, and Great Horned Owls. These threats were compounded by the Spotted Owl's slow breeding rate. The southern Mexican race, he explained, was somewhat more secure, given that its primary habitat was less susceptible to logging.

A pair of red-tails circled overhead, quartering the canyon. Smitty squinted at them through his silver-rimmed spectacles: "The owls will be higher up," he said, "they'll be avoiding the red-tails." We walked more slowly now, trying to attend both to the uneven, rocky path below and to the trees above where the owls were likely to be stationed.

We proceeded up the trail with the familiar stop-start rhythm of birders who like to walk, Smitty lingering each time we reached a known roosting area he wanted to scout out. I left the looking to his expert eyes. I was just enjoying the exertion and the distinctive aura of the Madrean pine-oak woodland that would draw me back to these border mountains again and again in the coming years.

I would be drawn back in particular by the vertical variety of the biomes in these parts. It was in these mountain ranges partitioning Arizona from Mexico that C. Hart Merriam came up with an essential building block of modern ecological thought: the life zone. Through his early twentieth-century investigations Merriam developed maps for these border mountains that combined readings of temperature, elevation, and moisture. From these he deduced that ascending 1,000 feet was equivalent to traveling 300 miles north, with a correspondingly decisive set of ecosystem changes. The diverse life zones, as he dubbed them, existed along the border in a luminous state of vertical concentration, unsurpassed in the U.S.

I was attracted to exactly this sensation: how a two-hour trip from the Sonoran semidesert floor, on up through grassland, through oak-pine woodland, through aspen groves, until finally I entered a world wrapped in alpine fir and snow-draped spruce gave me the illusion of having traversed a continent, from Mexico to British Columbia, in half a morning. The Huachucas and their neighboring ranges—the Chiricahuas and the Santa Ritas—are sometimes called sky islands, an apt term for the fairy-tale way these desert-encircled peaks reach into the clouds. For birders, these island mountains offer a magical world at an angle to reality, such is their improbable ecological variety and plenitude.

Smitty and I seemed to be crossing some kind of biome boundary now. All the way along our ascent, we'd been accompanied by chattering Mexican Jays. Now, for the first time, their calls gave way to the deeper rasping of Steller's Jays, which prefer the higher altitudes.

We passed a pile of ochre boulders and drank briefly from a cool spring.

"They must be at the upper roosting area," Smitty announced, swerving onto the trail's right fork. We hadn't gone fifty yards before he stopped abruptly. He nodded with his head, no finger pointing.

I saw from his expression that he'd found the owls, although I couldn't as yet discern them. I entered that elongated moment of anticipatory ecstasy familiar to all birders, when someone in your party has spotted something unusual, and you know you're about to see it too.

And there they were. Huddled close together, perhaps twenty feet above us. Their fugitive pigments made the owls seem continuous with the sun-flecked bark: my eyes had to peel them off the tree. Two tobacco-hued birds, with white mottling on breast and belly. They had a peaceable aura. I had to remind myself I was looking at political kryptonite.

Their facial masks—those parabolic discs that steer sound toward the ear hollows—were paler than I'd expected. Unusually for owls, they had dark eyes, the darkness somehow softening their gaze, making them seem more forgiving of our intrusions. Most sleeping owls, when interrupted in the day, blink open their eyes and blaze at you with an orange or yellow fire.

Spotted Owls are hard to find and even harder to detect, given their cryptic coloration and diurnal stillness. This pair would have been impossible to locate except by someone who knew every inch of their patch or seduced them with a recording. Across the entirety of their range, from British Columbia to the Sierra Madre, only 15,000 Spotted Owls remain.

Yet the owls meant for me far more than the sum of their scarcity.

Sometimes the best bird experiences are defined less by a rare sighting than by a quality of presence, some sense of overall occasion that sets in motion memories of a particular landscape, a particular light, a particular choral effect, a particular hiking partner.

Years after my hike with Smitty, I was teaching at the University of Wisconsin in Madison when I learned that a Great Gray Owl was hanging out in the parking lot of a suburban microbrewery. Although I do not keep a life list, technically speaking this impressive owl would have been a lifer and cause for some excitement. So en route from my last afternoon class, I drove to the brewery and parked. I could tell straight off that the owl was there. A throng of thirty birders had formed a semicircle, bins raised in unison, with the massed attentiveness of safari-goers bused in to gawk at a lion kill.

I got out of my car, hiked twenty feet, then stopped. The Great Gray balanced unsteadily on the brewery sign above a wedge of yellowing snow. The bird had a melancholy, remorseful look, as if it had intended to drop in for a single disciplined pint and had ended up imbibing way too many Spotted Cows.

I got back in my car and continued my commute, knowing I'd seen a Great Gray Owl, but that, stripped of any ecological ambience, it had been a theoretical sighting. The owl, in and of itself, proved an insufficient destination.

I have a visceral dislike for birding from a car. They say sitting is the new smoking, and my writing-teaching life requires that I sit to a dangerous degree. So I take any chance I get to stretch out my chair-cramped body, hiking for the hell of it, preferably in places where I might encounter some unusual birds.

My birding tastes were determined by my early teens. My passion and ornithological skill set peaked around age fourteen, when, alongside my younger brother, Andy, I hiked and hiked across the vast vistas of our childhood Eastern Cape, a region of South Africa replete with bird-rich convergence zones, places where the Indian Ocean shoreline, estuarine mudflats, montane forest, grassland, valley bushveld, semidesert, and the Cape floral kingdom's distinctive fynbos meet and mingle.

But by the time I was seventeen I had fallen out with nature in general and with birds specifically. The environmental idyll that had shaped my passions was, in retrospect, scarred by a segregationist ideology: we were white boys walking in the wretchedly divided world of apartheid. I fell out with birds and into politics, fleeing South Africa for political exile. I left in more ways than one, severing myself from all the pleasurable, now treacherous, memories that childhood birding had afforded me. It would take me some twenty years before I could return to the kind of simple enjoyment I could savor once more as Smitty and I beheld his Scheelite Canyon owls.

We lingered for half an hour, watching the owls do very little: preening, dozing, blinking mainly. Leaving them to their sleepy obliviousness, we moved back down the canyon. Smitty got talking about the threat posed by the territorial incursions of the more aggressive and more adaptable Barred Owls, which, assisted by climate change, were spreading westward rapidly and ousting their meeker, fastidious relatives. I began wondering what would become of this old-growth owl demanding an antiquated, almost quaint purity. The Spotted Owl's survival is impeded by a long list of nonnegotiable niche needs: dietary demands (red-backed voles, flying squirrels), nesting requirements, temperature range, and canopy density all highly specified.

It's a species ill-suited to the roiling world of the Anthropocene, where nimble generalists rule and hybridity is everything, a world of climate chaos, fractured habitats, and coming down the tubes ever more human perturbation. The unprecedented pace of anthropogenic impacts is creating novel ecologies that favor avian generalists: gulls, crows, Rock Doves, and Barred Owls that learn to flourish in the disturbed ecosystems that human activity generates. Will we be left with just the opportunists, the vulgar birds, like *Sturnus vulgaris*, the European Starling, which, from one continent to the next, adapts to every impossible circumstance thrown at it?

What kind of world are we bequeathing to unborn birders as our actions whittle away at ecological variety, at the earth's great multitudes? The Spotted Owl will never be a first adapter, a shift-shaping pioneer species, improvising a home out of a metropolitan garbage mountain or a second-growth forest logged to smithereens.

Midway through our descent, we heard a soft sound—a half-cawing, half-muffled grunting—coming from the canyon walls just ahead of us.

"Ever seen an Elegant Trogon?" Smitty asked.

I shook my head.

"Go."

And so I scrambled down the rocky trail, my clattering footsteps driving the trogon deeper into the canyon. Each time I moved, so did the sound, and I gave up on closing the ever-widening gap. I turned round to wait for Smitty only to discern, thirty feet away, a second, silent trogon halfway up a sycamore. A stout-bodied male, its emerald sheen set off exquisitely by its pink-washed belly.

One flick of the wings, and the trogon was gone.

But it lingered long after it had left. The trogon dropped me down the mine shaft of memory, where I remained for the rest of our hike. I'd seen a closely related species, thirty years before and 10,000 miles away, in South Africa's Eastern Cape.

When I was thirteen and my brother Andy, my perpetual birding buddy, was ten we'd happened upon a Narina Trogon—a species we'd never seen before—in a forest glade. The trogon was notoriously shy, which meant, primarily, that it was impossible to detect. With its emerald back to the viewer, the trogon merges with the luminously green subtropical foliage that it favors. It merges as seamlessly as any Spotted Owl absorbed by a thicket of old-growth oak. Andy and I, distracted by some Trumpeter Hornbills and Paradise Flycatchers, had overlooked the trogon at first. Then what we'd assumed was vegetation shifted ever so slightly, and we glimpsed a crimson breast. Almost before we knew what we'd seen, the bird shifted again and morphed back into pure forest.

My brother had been on my mind a lot these days. He'd recently reentered the outside world after long years in an asylum where he'd been institutionalized with schizophrenia. His transition out was rocky, to say the least. But he'd returned to our old haunts and found there—among birds and the people who were drawn to birds—a steadying community. Bird by bird, the stream of life he'd left was remade.

Birds had saved my beloved brother, now in his midthirties, had restored his sense of community and purpose, given him a shot at dignity. He knew his shit, as the birders in his patch soon learned. But he imparted his knowledge with an openhearted humility. He liked nothing better than passing on to new converts his well-informed enthusiasm.

On the phone we'd trade birding news, always seasonally inverted: I'd be reporting on my Snowy Owl sighting from the depths of the Wisconsin winter as he'd be regaling me with the flurry of summer migrants—White-tailed Bee-eaters, Alpine Swifts, Red-chested Cuckoos—he was encountering in the Eastern Cape. On a recent call, Andy had suddenly announced, apropos of nothing: "It was a fair exchange, you know. You gave me birds, I gave you books." It's true—a precocious book lover, he led me to the joys of immersive reading, while I, the older one, lit out first for the bush, blazing a trail for him. Books gave me a profession, birds gave him a lifeline after some very dark, shut-in years.

So that's really why I am up here in this canyon, in pursuit of owls and trogons and memories, in homage to my brother. The first person with whom I walked and walked and walked, listening, watching, waiting, alive to whatever might appear and alive to his perfectly companionable semi-silence.

Finally, binoculars to hand again, my brother was on the mend. All birds, everywhere, signaled that to me.

When we reached the trailhead, Smitty and I sat on a boulder, poured some coffee, took out our sandwiches.

I asked him how he'd first gotten into birding. It's a question I often ask of birders, hoping for something more personal than "I'm retired now and found I had extra time on my hands."

The question "why" can have the longest tail.

He moved a lot, Smitty said, from one military posting to another. Birds gave him an anchor, something steadying while shifting bases and countries all the time.

Then he fell silent, coughed into the pause, and reached farther back.

"Guadalcanal," he said. The word hung there for a while in the surrounding quiet. Somewhere, very far-off, I could hear a Canyon Wren launch its tumbling trill, no louder, from this distance, than a whisper.

In his early twenties, Smitty was dispatched to Guadalcanal, a remote Solomon Island that, between 1942 and 1943, witnessed a famously fierce set of naval, aerial, and ground-troop battles between American-led allies and the Japanese empire. But it was the aftermath of the American victory that left its mark on Smitty. The U.S. military tasked him with tracking down MIAs. So first in Guadalcanal, later in Papua New Guinea, he'd vanish into the dense, mountainous jungle, sometimes for weeks at a time, trudging through relentless tropical rain as he tried to track down the dead.

For company, he had some local Solomon Islanders who'd been conscripted (or more probably coerced) into carrying out the putrid bodies. But Smitty and the porters shared no language. His isolation felt complete. Even the moments of good fortune felt grim, when they discovered a dead man and had to machete their way through the jungle to ferry out the human remains.

Then Smitty started noticing, all around him, a plenitude of birds. Loud, luminous ones. He knew nothing about birds at the time, but they pulled in his gaze. They gave him a reason to look up, to look out, a reason to pay attention to some brighter possibility than another recovered corpse.

"So that's how my interest in birds began," he said, adjusting his silver-rimmed spectacles.

Later I would look up the Guadalcanal checklist and wonder what he saw. The Superb Fruit Dove? The Glossy Swiftlet? The Red-breasted Pygmy Parrot? The Pearly Owl? The Dollarbird?

As Smitty talked, I glimpsed a young man dispatched to the far end of the world and struggling to hold on, a young man suffocating beneath the stench of war. A young man whose survival instincts fostered in him a kind of double consciousness as he roamed the jungles of Guadalcanal, on the qui vive for corpses of his countrymen and for sheeny tropical birds.

My brother, at a similar age, had fought a different kind of war, under daily threat of hearing the thread of sanity snap that linked him to what passes as the normal world. And yet one thread remained, helping haul him back to someone who resembled the birder he'd once been, allowing him a tentative set of hopes and anticipations, glimpses of roseate color through the deadening green gloom.

I found myself returning in the days that followed to my window into Smitty's double consciousness. The upper reaches of Ramsey Canyon proved to be a place of liminal complexity, not just ecologically but politically as well. In my wanderings, I happened, again and again, upon heaped clothing and canned goods jettisoned by Mexicans crossing over into Arizona in the dead of night. With each step I took, the trail beneath my feet released the soft scent of fallen leaves, but hanging in the air was another more ghostly odor, the scent of migrant fear.

I started to draw a line down the middle of my spiral notebook. In the left-hand column I noted the bird species I encountered. On the right I noted the objects abandoned by the would-be immigrants—the things they'd carried—as they descended from the sky islands and started to shed their load in preparation for the desert ahead. By week's end I had a notebook that looked like this:

Magnificent Hummingbird	Gatorade (Lima-Limon)
Ferruginous Hawk	Electrolit/Suero Rehidrante (para prevenir o tartar la deshidratacion)
Ladder-backed Woodpecker	Tortillas de Maiz
Blue-gray Gnatcatcher	Tempra paracetamol (olor de cabeza ocaionado por estres)
Cassin's Kingbird	Herdez Granos de Elote [sweet corn]
Verdin	Mazatun (Atun Aleta Amarilla en aceite) [tuna in oil]
Prairie Falcon	Single, torn tennis shoe
Scaled Quail	Used diaper
Curve-billed Thrasher	Bimbo Panque con Nuez [walnut cake]
Unidentified species of *Empidonax* Flycatcher	Unidentified species of candy bar wrapper

Sometimes you embark on a pilgrimage without being aware that you're a pilgrim. I'd thought Smitty's suggestion that we look for owls on O.J. day was a spontaneous, creative act of generosity. Generous it was. But many years later, thinking back on the poised pleasures of our hike, I looked up Smitty on the Internet. To my surprise I learned that since discovering the Scheelite Spotted Owls in the mid-1970s, he had led countless seekers to them, and would continue to do so until his death in 1998.

"Countless" is inexact, a word Smitty would have shunned. He was big on counting. He recorded and ordered the details of every bird trip, as if his life depended on it, which it did in a way during his Guadalcanal days when he went questing for the dead. To the last he documented, with actuarial precision, the time of day, the weather, the precise location and behavior of the Scheelite owls, the names and numbers of any predators present, and the names of the people he was guiding. He prided himself on being able to deliver on his promise, on having, in this ritual quest, a 97 percent success rate. All in all, he led over 6,000 birders to his owls, never in groups larger than twelve, more often in ones and twos.

So there I was avoiding the hoopla of the O.J. trial while being steered, unbeknownst to me, in the direction of celebrity owls.

What I remember most from that day with Smitty is the clear joy of birding with someone who had a similar conversational gait, someone who, as we walked, conveyed a companionable silence, punctuated by bird-sensitive, restrained exchanges. I was lucky enough in life to find that quality in my younger brother early on and have continued to seek it out. In the nine-person, four-generation household where I grew up, a kind of chaos ruled. Together my brother and I sought to escape the mayhem and improvise our own variety of order through interminable bird lists. Birding sustained but also distorted us, camouflaging an unexamined pain. Yet later our ornithological precociousness would become a great well from which we both drew solace. Now wherever I encountered birds, I felt accompanied by Andy, a man whose disordered mind had rediscovered the power of well-ordered observation. And rediscovered all the community—the human and avian community—that flows from that.

Chapter 3

Birding in Traffic

Jonathan Rosen

On a cold morning late last month, I took a subway to Union Square Park to see a bird I had never seen. The bird, a Scott's oriole, had been noted intermittently behind the statue of Mohandas Gandhi since December, though it took birders several weeks to figure out that it was not in fact an orchard oriole—which would have been unusual enough for winter in Manhattan. Scott's oriole is a bird of the Southwest and has never been recorded in New York. It should be no farther east than Texas, which is why, despite my sluggardly winter ways, I decided it was worth a trip down from the Upper West Side, where I live.

Alongside my excitement, I felt a qualm of embarrassment as I exited the busy subway with my binoculars. It was like taking a taxi to hunt big game: "Let me off near the wildebeest, driver." In Central Park, I can at least conjure the illusion of wildness if I focus on the trees. But when your marker is a metal statue of a man in a loincloth, standing on what is essentially a traffic island, you cannot pretend you are in the middle of nature.

Then again, that's the point of bird-watching. "Nature" isn't necessarily elsewhere. It is the person holding the binoculars, as much as the bird in the tree, and it is the intersection of these two creatures, with technology bringing us closer than we have ever been to the very thing technology has driven from our midst. And, anyway, there are still wild elements in the center of a city. The morning I arrived, the bird had made itself scarce, perhaps because a red-tailed hawk, a Cooper's hawk, and a kestrel were all patrolling the park.

I was not the only birder there. Everyone had read the same birding e-mail messages I had, and we were all staking out the southwest corner of the park, scanning the same stunted holly trees and viburnum.

Oranges and banana slices had been scattered on the ground, like votive offerings. The first report I read of the bird had it eating a kaiser roll. Several people had been there for hours, and two men showed me pictures of the bird that they had taken on their digital cameras that very day. They were hoping for a last look and braving the cold in the knowledge that by noon, sunlight would again fall on the building-shadowed corner of the park and entice the lemon-yellow, black-headed bird back into view.

Vagrant though the bird was, it seemed to me that there was also a rightness to its having landed in Union Square. This was not simply because of the statue of Gandhi, suggesting the need for simplicity and putting me to shame in his cotton dhoti and sandals as I shivered in my down jacket. My feelings also had to do with the park itself, named originally for the union of Broadway (then called Bloomingdale Road) and the Bowery.

Bird-watching is all about the coming together of disparate things, not merely earth and sky but the union of technology and a hunger for the wild world. "Imaginary gardens with real toads" is how Marianne Moore described poetry.[1] Birding in city parks evokes much the same sensation. The parks, and the cities around them, may be human-made, but the wildlife that flashes through is no less real.

On the building across from where I stood, high up on the brick wall, there was a metal box that from time to time emitted the cry of a peregrine falcon. It was just a recording, but it roused the pigeons on the windowsills into a sort of lazy panic, getting them to rise and fly a few circles in the air before resettling. Even real peregrine falcons have a hint of the artificial about them, having been brought back from the brink by falconers expert in the ways of an ancient art that involved borrowing a bird from the wild and then turning it loose again.

Like the greenmarket in Union Square that brings apples and vegetables from outside the city, the token bird in the park is a reminder of an older way of life we are still intimately connected to and vitally in need of.

And like birders with their binoculars, we are not necessarily doomed by our modernity to exclusion from wildness. Bird-watching was born in cities—combining technology, urban institutions of higher learning, an awareness of the vanishing wild places of the earth and a desire to welcome what is left of the wild back into our world.

The name Union Square accumulated layers of later meaning, from the great rally held there in 1861 for the Union troops, and the Labor Day marches that took place later that century. In its own way, Scott's oriole belongs with Union Square's famous 19th-century monuments, most especially the 1868 statue of Abraham Lincoln.

The bird was named by Darius Nash Couch, a Union general who was also a naturalist. (There were a lot of army men in the 19th century who used their postings as a way to record bird life.) Couch named the bird in honor of Gen. Winfield Scott, who was known as "Old fuss and feathers," though I feel sure that is not the reason he got a bird named after him; one of the great soldiers in American history, Scott began his career with the War of 1812 and ended it with the Civil War.

The bird is a monument to nineteenth-century ornithology, but it had defied its label and was doing what creatures with wings do: flying out of range and surprising us with life. It is never enough to know the name of a bird when you are birding. It is the mysterious unknowable animal that lives alongside the named and classified creature that draws us out to look.

By noon on that cold January day, about twenty birders had gathered, craning with increasing urgency into the bushes as the little patch of grass behind Gandhi grew brighter. And suddenly the bird was there. Someone pointed, and then we all saw it.

It came down to the ground and, without ceremony, pecked at a piece of banana.

Chapter 4

Buried Birds

Elizabeth Bradfield

Standing on a boat with no land on the horizon, there is no other word than love to accurately describe what surges through me when I see a Sooty Shearwater gliding stiff-winged over the waves. The way they tilt in flight, each wing tip in turn nearly cutting into the glassy, reflective surface, reminds me of kids playing airplane, holding their arms out to either side, running and banking. It reminds me of the freedom a child-body has—my child-body had. The delight of wind's rush my own speed created, the almost dizzying sway, the expansive sense of loft, the unselfconsciousness. I remember actually feeling that I *was* flying, that my imitation somehow embodied that freedom. When a shearwater glides by, all those sensations echo up, and they are undergirded by my adult knowledge of these birds' annual record-shattering, looping path around the Pacific Ocean basin, their phenomenal endurance and also their fragility in a changing, plastic-saturated world. But the thrum in me is something deeper than intellectual admiration, concern, or appreciation. It's visceral.

Not only the shearwaters move me. Other pelagic species do the same.

Working as a naturalist aboard a small boat in the Gulf of California one winter, after hours on deck staring at water, I suddenly spotted a Xantus's Murrelet. Even though I'd never before seen one, the squat, compact body and two-tone coloration were unmistakable. A murrelet! This far south! I knew that we were in their range, but in my mind alcids are inextricably associated with the chill of northern waters. For years I'd hoped to see one there, in Baja, and now the very thing was before me.

"Alcid!" I shrieked in a high, uncool, and very over-excited voice.

Everyone on the bow startled, turned and looked at me. Carlos began laughing, but I didn't care. I was pumped through with adrenaline. Heart racing, pectoral muscles tingling, I felt like I'd erupted with (been erupted by?) joy. Is this what being struck by lightning (when it's not fatal) is like? It seemed that shocking, that bright and electrified.

I used to think that I loved pelagic birds—petrels, razorbills, albatross—because they spend most of their lives at sea. Because many, such as the albatross, can sleep on the wing. Or because those like the murre can fly to great depths underwater.

All of that is true. And yet it's not only the true pelagics that I love.

It's the gulls, terns, and even the "shag rats" (a local name for Double-crested Cormorants) and boobies. I am glad to tease through the early years of gull plumage to figure out species, an exercise akin to puzzling through Encyclopedia Brown mysteries, as I did when I was a girl. I am glad to wince at the sharp tang of a rookery as it hits my nose.

Jaegers dogfighting a tern? Yes, please.

I love sea birds because of their differences from "regular" land-based birds. Because most have long-term pair bonds, long lives, shared parenting duties, and little sexual dimorphism. They spend much of the year apart, then come together for an intense period to (hopefully) rear a chick or two in a dense community of their kind. They know both water and air. They travel vast, open distances. The option of a different conventionality—another way of life—first appealed to me as a young queer woman who knew that the examples she saw around her would not guide her. These different birds were beautiful. There could be a different way of thriving.

But above and beyond even all these details, my heart is held most fully by the many species of pelagic birds who hide their lives on land. The ones like prions who bury themselves to lay an egg, whose chicks hunker in the dark until fledging or trundle out to sea before they can even fly. These birds know how dangerous our element—earth—can be.

I never tire of watching them or imagining their presence when they can't be seen.

The Blinds Must Be Closed by Dusk

In for the night. Porthole a raven's eye glinting back the lit lamp. Forgot before dinner to shade it. From bird's eye, outside, this glass disc I look to & through appears a moon which could but didn't this time tempt petrel or prion, birds that rush from burrow to wave. Know the danger of transitions. Are moth-like toward light. But all the same and despite my forgetting once again no morning call to release a thing towel-wrapped & box-burrowed by A on night watch.

illicit hope:
one flumbles to deck lights &
then, savior, I hold one

Nothing lured. No memory of anything leaving my hands which then would be light, warm, & perhaps, for a flitch of time, feathered.[1]

Some of my most vivid memories of birds are not my own. They are my wished-for memories.

One is the story about storm petrels coming into their burrows at night told by Richard Nelson in *The Island Within*. Nelson weaves his story with the 1896 experiences of Joseph Grinnell on an island he calls Kanaashi in the temperate rainforest of Southeast Alaska (or British Columbia? He is wonderfully elusive in his pinpointing).

Near midnight, Nelson writes, "the air is swarming with petrels, like feathers falling in a whirlwind. I can still pick out individual birds against the dimly lit sky, darting back and forth, swift as hunting hawks, delicate as dancing butterflies, agile as looping swallows. If their wings make any noise, it's lost in the clamor of calls."[2]

Envy, awe, and an almost sexual ache of longing fill me.

I want that.

I've seen specimens of storm petrels that sailors (and others) would use as candles, a wick threaded down into their oily bodies, string lolling like a strange tongue from the open bill. What did it smell like, I wonder? The stomach oil of those birds, the burning of its own dried flesh? It seems both beautiful and barbaric. A candle from the body of a bird, a bird just the right size to fit into a fist, its feathers warming the hand that holds it.

I've seen the birds Nelson mentions, the island he mentions (or one like it). I've walked the shore, peered into the valleys riddled with burrows without seeing a single bird. But I've never been there at dusk, flashlight clenched and covered so as not to disturb, my hand hot from the bulb but also with the feverish wish for light, for what the light might call toward me. I've never slept in a tent there, birds hitting the rain fly throughout the night, calls coming up from the burrows.

And I've never been there alone. Always with a group of people for whom I'm guide. So, in a way, I haven't been there at all. I haven't seen them at all.

But wait. Just this week, working on a ship in the Aleutians, I was called out of my bunk at midnight by one of the crew. There was a bird trapped in the pool, she said. D and I pulled on boots and jackets, grabbed a towel, and rushed out on deck. It was a Leach's Storm-Petrel fluttering and stuck beneath the netting tied over the empty swimming pool (in all my time aboard, I'd only seen it filled once). Against the turquoise tiles lit by deck lights, it looked even more delicate than when seen fluttering above the waves. D climbed down a ladder and walked slowly, holding the towel before her to trap the panicked bird. An ornithologist, she'd worked with petrels in the North Atlantic for many years. She checked the wings (fine and fine) then lifted the bird to me. I reached out my hand, first and second finger wide in a "v" to hold the neck. Its small bill tapped my knuckle. My thumb and fingers curled around its soft, warm body, claw-like, trapping but not gripping. We walked it down to the fantail, the darkest part of the ship, and I set my other palm beneath its sweet, leathery, webbed feet, opened my hand, and it flew off, tilting over the wake in its typical, butterfly-like flight.

D brought her fingers to her nose. "Smell that? I love the smell of petrels," she said. I lifted my hand to my face. Just my own skin-smell. No evidence at all to breathe in. Then D held her hand to my nose. Musty. Soft. A rich, subtle perfume.

Muttonbirds, Port Fairy, Australia (January 1, 2016)

> *Bridge to island over water that could be waded. Island fortified, pathed, prepared for threat by sea. Inland, confusion of succulents overgrowing each other over dune-mounds, green and loose. Baby joggers, track suits, kids on bikes circle the paved perimeter, take the lighthouse spur. Ibis preen. Yellow wattle of masked lapwings bright against purple heather. But what holds attention, imagination is shadow, shadow dug into white sand shadowed by karkalla, the carpet of sound it will become in darkness.*
>
> *silent tunneled dunes*
> *hunched in dark depth, unsleeping*
> *one and one and one*
>
> *Peer in, hopeful. Imagined eye. At one entrance, unfleshed wings, splayed. Adult taken by hawk or rat. Dear unfledged (unhatched?) one, I'll look for you off the Aleutians in August.*[3]

Another unlived story that shines in my memory is one related by a friend. She heard it, in turn, from a friend. It takes place on Penikese Island, a rocky hunch in Buzzard's Bay, Massachusetts.

Penikese has a fascinating history. From 1873 to 1875, it was the site of a research institution that would later become Woods Hole Marine Biological Library. Then it was a leper colony—the only one in New England. I can imagine those poor people, islanded, isolated but in view of the mainland. It must have been so cold and damp in the winter. The first doctor to the colony quit, not able to face the solitude. After the colony was disbanded in 1921, the state razed the buildings and left Penikese as a bird sanctuary. Fear of contagion prevented any other use . . . but the story doesn't end there.

I have been on Penikese only once, back in the late 1990s, when I took a boat to look for seals as part of an Audubon day trip. It was a cold time of year, most likely early spring, though I can't really remember what month it was. We landed on the island and walked around. Orange-gold lichen splayed across the large stones of the old buildings. Someone gave us a tour of the facility, which, since the 1970s, had been repurposed to serve as a rehabilitation school for young men who had broken minor laws.

Penikese is low and rocky. Small. Only seventy-five acres. Yet there are spaces to hide. Swales and nooks and those crumbling, loose-mortared foundations.

Here's the story my friend told me: in 1972, a boy on the island heard a strange sound in the night. Something different than a mouse, which no doubt he often heard skittering in the dark. I wonder why he was awake. Was it bullying from the other boys? Fear at the absence of urban sounds? Just ordinary restlessness? He wondered what it was.

Some of the story details are fuzzy to me. But at some point the boy told one of his schoolteachers what he'd heard. I can't quite believe that he was lucky enough—that we are lucky enough—the teacher listened to him and was curious, too. The next night, both stayed up alert, attuned.

It turned out that they had discovered Manx Shearwaters nesting in Massachusetts.

Manx—quick-flapping and stiff-winged, smart black and white plumage. This species wasn't supposed to be on Penikese. They weren't supposed to be breeding in North America at all. Most of the world's population still breeds on islands off Wales, Scotland, and Ireland. But there they were, on Penikese, discovered by a kid almost written off, a kid isolated "for his own good" on the kind of refuge island these birds seek.

I don't know what it's like to be held at a boys' school for rehabilitation. I don't know what it's like to discover a nesting bird. But my young self knew isolation, confusion, and wonder. I wish I'd been given the gift, then, of birds.

Prions, New Island, Falkland Islands (December 25, 2011)

White sand hard packed, beach crescent oystercatcher-stippled.
Copper-sheathed ship stem sculptural, wrecked gentle. Perfect porcelain commode visible through hull hole, bright and clean and tilted. Useless.
Stroll the island's saddle to bird amphitheater: albatross, rockhopper, shag.
Skua working to down a downy chick headfirst, webbed feet flopping.
In landing-prep read the fact sheet & am lost still at what is unseen and surrounding and nearly underfoot everywhere—

in tunneled tussock
millions of silent prions
hunker until dusk

Favorite bird, hard to see even at sea, wave-blue, white-edged, an unreality fleetingly real tilting above our wake beyond sight of land.[4]

I did not grow up a birdwatcher. In my growing-up world there were robins, ducks, seagulls, eagles, crows, and geese. Ducks, to be clear, were Mallards (did any other kind exist?). Sometimes, we'd toss bread into a pond in order to watch them come to us. This was less about the birds, somehow, and more about the power of being able to dispense bounty or incite a frenzy. The birds were aggressive, rustly and sharp-beaked, poking the sodden crumbs with quick strikes. Standing by the water's edge, a sweaty clutch of plastic bread bag in hand, I looked down on them, I remember, with a mix of disdain and fear.

Other birds lived in my Pacific Northwest suburban childhood, but in our family we did not notice them. Maybe a hummingbird now and then. Maybe a jay or the hammering of a Pileated Woodpecker. My grandmother kept birdfeeders on her porch and a pair of binoculars on the kitchen table to watch things flit in and out. This seemed odd to me back then, an activity of an older and perhaps old-fashioned generation. I remember the binoculars had a tan leather case and a cracked leather strap. They were heavy and black and when I peered through them, I couldn't get them to focus very well.

So, what changed? How did I become a birder?

Whales were my gateway animal. And how did I get to whales? From boats. From a love of boats. Just out of college, my literature degree in hand, I signed on for a year as a deckhand on a boat that would travel from Alaska to Baja, showing tourists the sights. In that year, a few early signs of my path toward true bird-love can be found, moments of attention that were held: Tufted Puffins pounding their bright orange feet over the surface of the water as they ran toward flight, yellow breeding-plumes trailing behind like bicycle handlebar streamers (but everybody adores puffins, I was not unique in that delight); a spiral of sound that ascended through a glacial valley (Swainson's Thrush), but that was more atmospheric than specific to the bird itself; the slight shock at first noticing red flesh around the yellow eye of a Black Oystercatcher, so dinosaur-like and intense. These moments felt like part of a larger ambience of wonder, though, not a goal of attending to birds in and of themselves.

In 2000, at age twenty-nine, I began working on Cape Cod whale watch boats out of Provincetown as a naturalist. The year before I'd been a kayak guide, and had spent my summer on the water, too, but this was different. On the whale watches, we'd head out for our three-hour foray, sometimes leaving land far behind. We were in a different world, the world I'd been missing since I worked and lived aboard boats on the West Coast.

At first it was a scramble to just know what to say—the safety stuff, the crew intros, the basics of geography and geology and whale biology. I wanted to get it right. And when we did find whales, I honestly was agog with looking. Their huge flukes, the way water shifted flat immediately before they rose to the surface, the pink inside of the roofs of their mouths when they came up, maws open. I felt so far from "guide" or "expert" that it was ridiculous. But watching whales involves a lot of time on the water *not* watching whales; watching the water and waiting for whales. And when the whales were up, there were these noisy fluttering things above them sometimes, too. Often, they'd lead us to feeding whales. I started to pay attention to what moved through air.

"Birds," the captain would say, and we'd see a loose cloud gathering over the water and steer toward it. And then the dark water lightened, the green glow began, the whale huffed up.

When did "birds" become interesting beyond just being a mere advance team or early warning system? I am not sure I can really say. Sometime in that first year? When I learned enough about cetaceans to feel comfortable being loose with my presentation and got curious about the "other stuff?" Probably.

Although the linkage is strange, it seems important to say that Provincetown was the first place I claimed as a chosen home. I didn't grow up here or move here for a job—I chose it. And that choosing was a joyous and fierce claim. It was a remaking: leaving the west, leaving a dot-com job for the uncertainty of life as a writer and naturalist, dedicating myself to not only enjoying "nature" but also learning about it—plants, tracks, cicadas, salamanders, warblers—from amazing self-taught naturalists, making a home for the first time with someone I loved, living in a place where queerness was in most cases the default assumption. This home was a discovery, a place that freed me.

I remember one season early in my time on the waters off Cape Cod. A glossy, glassy day in summer. The air a bit soft and hazy with humidity. Light swell from the east. Hundreds upon hundreds of Wilson's Storm-Petrels fluttering over the water in their distinctive, wobbly, flittery way. It looked like pepper had spilled on the ocean. Like water-striders on a vast pond. Their thin, foolishly long legs dangled as if they were the puppeteers and puppets all at once, jerking themselves up and down above the surface, beaks nipping for plankton.

Someone told me that petrels had a sense of smell. I hadn't thought about birds and smell. Someone told me they may well be the most numerous bird on the planet, and yet I'd never seen them until now. What else could I not know? What else? A hunger began to grow.

A fluttering of storm petrels. The seeming vulnerability of their tiny bodies engenders an upwelling of emotion similar to that evoked when I at last see my sweet niece and nephew who live across the country from me. Overwhelming, maternal, deeply embodied. Full of closeness and distance, joy and sadness, at once. What a gift.

Atlantic Puffins, Lunga Island, United Kingdom (May 25, 2015)

Delicious, delicious, delicious this wild alongside the long-storied shore of Scotland. Ferry guests from ship through the ebbing channel to rough rock notch. Blue towels laid over barnacles & algae as welcome carpet and (hopeful) slip-prevention. Job swap. Hand off my boat and hike loose cobble, scramble thrift-tufted turf-slope to scarp-top, a lush, green ledge. Join the throng, both gore-texed and feathered.

busy busy
hop, bow, pluck greens, preen, posture
hushed, tunneled eggs below

Hundreds. Thousands. Hatchet-bills and orange feet bright. We sit, stare, and no one crowds them. Amazingly, I don't have to scold one encroacher. A huge-eyed rabbit works its jaw and I remind myself that it is not introduced, not invasive, but belongs.[5]

Yesterday, June 14, I saw my first Cory's Shearwaters of the 2016 season. They came across the Atlantic Ocean to Cape Cod from Algeria or Portugal or the Azores. These are non-breeding birds. Maybe they failed this year, either not able to reunite with a mate who was lost at sea or their single egg plundered by raptor or rat. Maybe they're young and not yet ready to commit to the annual egg-laying, incubation, and chick feeding that will occupy so much of their long lives. It used to be rare to see Cory's here, but in the last few years, there have been so many. I don't know why.

My gladness in their bright bills, their strong-winged soaring flips toward questions. Are we seeing more because of a greater range in juveniles? Greater instances of nesting failure? Is this a sign of woe?

Pelagic birds are sentinels, bringing back news gathered over the thousands of miles they travel before returning to land. Their presence is a connection to another place. For me, their presence is also an emotional door to other connections. In the deep watching of creatures, I drift through thoughts of friends (J's sick father, the awkward conversation with M), family (what would my grandmother think of the light this afternoon, this gannet's dramatic plunge? how I wish I could show her), politics (can the Dutch really invent a technique to deal with the ocean's plastic gyres?), snippets of song, facts about the birds' life history. It all becomes loose, blurred and blended by the immediacy of a bill or the cacophony of overlayered calls. Attending birds at sea is at once a respite from the scatter of land-based obligations (no cell signal, no other errands) and also a time when my mind can move as they do—soar, dip, sheer, glide.

But what *is* it about these land-avoiders that so deeply moves me? What is it that gets me about these birds who would spend their lives at sea if only they could find a place to put an egg on the waves? We resonate with certain animals, I believe, because they are a physical embodiment of an answer we are seeking. A sense of ourselves in the world that is nearly inexpressible.

Does it have to do with my need to be a secret poet among biologists, a secret naturalist among poets? These birds are secret land-dwellers at sea, sea-goers on land. It's true that I feed on the dynamic swing between aspects of my life. I need the community shearwaters have in nesting, the solitude they have in ranging. I need the time working on boats as part of a crew, the time alone in a chair adrift with words.

Pelagic birds have long incubation times and do not easily recover when one egg is lost. I am slow. I am slow to decide, to determine, to recover from hurt.

Or does it have to do with my own queerness and sense of how that has shaped me in the world? The lives of egg-burying, colonial birds like petrels are such a clear expression that dangers lurk, that community is key to survival, that expressions of gender can be hard to read for outsiders. The way they develop in darkness and emerge to light, born in a space that is not suited for a full life and then (hopefully) emerging to the realm in which they can soar . . .

The life of these birds is so different, so beautifully *other*. I suppose you could say the same for anyone's loved species, but these are mine. And so I connect my loves to my love of them.

Chapter 5

The Problem with Pretty Birds

Andrew Furman

The problem with pretty birds is that they are so hard to ignore.

There we were in our breakfast nook, my wife and I, assailing each other over our oatmeal with our respective workplace obligations, which ought to excuse us from competing childcare duties this afternoon. We hurled the important words of our important professions like stones across the breakfast table. *Mediation. Office hours. Deposition.* I knew pretty early on that I was going to lose this particular battle, mediations and depositions (whatever the heck *they* are, exactly) taking precedence over office hours, which I could cancel. But I wasn't ready to give up so soon. It was a bad mood that I was nursing, which I intended to nurse for at least a few more aggrieved sentences.

But then a painter's palette with wings over my wife's shoulder flashed against the sun outside the glass door, uttering silent sentences of its own. *Here I am!* it cried, alighting on our bird feeder, jutting its cherry chest and throat. *Here I am!* it cried, pivoting on its perch, showing off its emerald backpack now, munching millet between its mighty bunting mandibles. *Here I am!* it cried, dipping its whole royal blue head back inside the feeder's mouth for more millet, seed-hulls flowing from its beak like something molten as it emerged. A male Painted Bunting, first of the season. Around this time each year, late September, these birds abandon their twiggy, grassy, leafy, cobwebby, horse-hairy, and rootlety nests in north Florida, the Carolinas and Georgia and stay with us in the warmer subtropics until mid-April or so. So we've sort of been expecting him. Yet not now. Not now-now, in the middle of our domestic spat. I wasn't feeling cherry, or emerald, or royal blue. For crying out loud!

Have they no sense of occasion, these Painted Buntings? The answer, of course, is no. They don't care a fig about us or our moods, which is another problem with pretty birds. It's nicer to imagine, in the spirit of Ralph Waldo Emerson, perfect sympathy between the realm of nonhuman nature and us. Wallace Stevens, however, hews closer to the truth about birds and people in his poem, "Of Mere Being," when he writes, "A gold-feathered bird / Sings in the palm, without human meaning, / Without human feeling, a foreign song."[1] Though we eavesdrop, shamelessly, the birds don't sing for us. Our relationships with them, and with most wild creatures, are terribly one-sided. Hardly relationships at all.

Still, it's not like I could exactly ignore this pretty, problematic bird outside, over my wife's shoulder.

"There's a Painted Bunting at the feeder," I said sharply, joylessly, as if to say, *I'm angry at you*. Which is what my wife might actually have heard, as she continued:

"You know I can't cancel the mediation, honey. My clients are flying in from Omaha."

"Do you fucking hear what I'm saying, Wendy!? It's a goddamn Painted Bunting for Christ's sake! At the feeder!"

"A male?" she asked, nonplussed, finally hearing the key words through my ludicrous volume, tone, and timbre. She pivoted in her seat, glanced over her shoulder out the glass door. "Oh, he's so pretty," she said, as if to say, *Oh, he's so pretty*, adjusting more seamlessly than I was able or willing to adjust to the morning's shiny new terms.

The thing is, it's not merely pretty, the Painted Bunting. It's outlandishly, ludicrously, ridiculously pretty. "The most gaudily colored North American songbird," Roger Tory Peterson writes.[2] *Nonpareil*, the French name for the bird, "without equal." Its blue head somehow bluer than blue. Its green back "electric," opines Peterson. The chest and neck not a mere red or even cherry. Vermillion, rather. And all three of these colors on the same small bird! Colors so vibrant that the winged creatures do seem electrically enhanced. A Christmas-light bird. *Look here!* male buntings seem to say. *This is what blue and green and red ought to look like!*

Hard to fathom that such a bird has evolved over millennia, existed, and exists, alongside scruffier sparrows and finches and flycatchers in North America, alongside scruffier us. A male Painted Bunting makes you wonder, if you're the wondering type: why this particular, improbable animal form? Why these bold contrasts in hue? Why emerald green here, royal blue there, vermillion here? More ordinary, extraordinary curiosities arise, while you're in a thinking mood: This beak? These wings? These spindly legs and tiny claws? What strange and wondrous forces issue such a creature into being?

Moments like these, when a pretty bird interrupts an irascible mood, I'm reminded of how poor a watcher I truly am, or have become in my harried adulthood. The greater patience of other writers frequently puts me in my place. Like Annie Dillard, who summons spectacular imaginative resources in *Pilgrim at Tinker Creek* (1974) to engage with the natural world ever more mindfully. "When I lose interest in a given bird," she writes in the "Spring" chapter of her classic, "I try to renew it by looking at the bird in either of two ways. I imagine neutrinos passing through its feathers and into its heart and lungs, or I reverse its evolution and imagine it as a lizard. I see its scaled legs and that naked ring around a shiny eye; I shrink and deplume its feathers to lizard scales, unhorn its lipless mouth, and set it stalking dragonflies, cool-eyed, under a palmetto."[3]

The male Painted Bunting sports a naked ring around its eye too, a crimson contrast against its royal blue head. Rarely, however, do I look at these birds concertedly enough to really notice this crimson ring. There's the person that we are and the person we'd like to be, and the best we can probably do in this life is nudge ourselves, through conscious Dillard-like effort, ever closer to the latter. The other thing we might do is adjust our expectations for ideal selfhood every once in a while, as I've done (and as the preservation of one's sanity dictates). But I still feel that it would behoove me to exercise more patience, more mindfulness, before the actual outdoor world. I doubt that I'll ever match Dillard's patience—or the patience of so many other writers whom I admire, past and present—yet I can surely do better.

And so . . .

Wendy and I rose from our seats at the table, stood before the glass door and watched the pretty, problematic bird, outside. What else could we do in the presence of such a visitor? I called our four year-old daughter, Eva, over from her puzzle on the family-room rug to glimpse the Painted Bunting, too.

"You see it?" I asked, hoping that her spongy brain would absorb the image before it flitted off into dense cover. She's just at the age when memories begin to stick. Wouldn't it be nice if she were able to summon, years from now, this fleeting, feathered vision?

"I see it," she uttered, nose to the glass, pleased but undazzled. She watched the bird for a few moments, then skittered past my knees back to her puzzle. Okay, maybe. Okay, that for now she felt that it was perfectly ordinary and unremarkable that she shared a world with these bejeweled birds.

Three female buntings—now four!—emerged from the nearby firebush and necklace pod foliage to join the male at the feeder. Pretty in their own way, these females, green stem to stern, a bit darker-dashed here and there, as if these few feathers were dipped in water. If these green and dark-dashed birds were the male Painted Buntings, say, and female Painted Buntings were a drab brown, all we'd talk about was the beauty of these small green and dark-dashed birds. But these aren't the male Painted Buntings so no one talks about the prettiness of plain old green and dark-dashed buntingness.

Five females at the feeder now and still this single male. His harem? Why is it, I wondered, that we always see so many females and so few males each year?

The problem with pretty birds is that that they tend to get eaten by other birds. Cooper's Hawks, Sharp-shinned Hawks, Peregrine Falcons, Merlins. All of whom seem to make a decent living here. Solitary, pugnacious killers. The Merlins, especially. I saw one just the other day in coastal scrublands near my home, perched atop a withered sand pine surveying its domain, silencing the nervous warblers and vireos in the canopy below, its slate-gray back and speckled chest puffed up against the salt wind.

It may be that male Painted Buntings, who surely winter here in equal numbers to the females, are simply more skittish and covert than female birds, given their outlandish, ludicrous, ridiculous prettiness. Their feathers, after all, simultaneously shout *Love me, love me, love me!* to female Painted Buntings and *Eat me, eat me, eat me!* to most everything else, including Merlins, including (come to think of it) the ever-expanding band of feral cats in my neighborhood, which rove about most suburban neighborhoods these days.

The problem with pretty birds is that they tend to get trapped and sold by resourceful humans, too. Easy to lure inside wooden cages with "rival" decoys, the ornery Painted Bunting males. Thousands caught every spring, observed John James Audubon in 1841, shipped from New Orleans to France, where they fetched a handsome price. Still taken in large numbers in Mexico, Central America, and the Caribbean, sold at flea markets, some for the cage-bird market, some to compete against one another in underground singing battles. I suppose that such clandestine events are somewhat analogous to cockfights or dogfights, but it's tough for me to imagine these gatherings in quite the same light. A clan of human malfeasants, drinking and smoking and gambling over the singing prowess of pretty birds? Is it possible that a more formal air perfumes such contests, that men and women don their Sunday finest to listen to the sweet warbles of Painted Bunting competitors?

Pretty birds, provided they don't get eaten by raptors or feral cats, or trapped by nefarious humans, entice mates with greater success, thereby increasing their reproductive fitness. You see the tension. Clearly a balance must be struck between these competing interests. Enticing mates. Eluding predators. Most finches and sparrows seem to have it figured out pretty well. Earth tones. A few stripes here and there, a swatch of color maybe at the crown, lores, or eyebrows. Nothing crazy.

Not so the Painted Bunting.

It might have been a good idea for Painted Buntings, before people were around to call them Painted Buntings, to have convened a Council of Learned Elders within the cover of greenbriar or myrtle. A male bunting might have gazed out at his cohorts across the latticework of branches, offered a proposition to the females, *Listen, we know you like pretty greens and blues and reds, but we live in a world with Cooper's Hawks, Sharp-shinned Hawks, Peregrine Falcons, and Merlins. So let's not get carried away.*

But as you say, a female bunting elder might have replied, *we do like our greens and blues and reds.*

Which might have elicited the following response from a separate male elder: *While we cower within these branches, look over at those Savannah Sparrows frolicking out in the open field there. They didn't get carried away, see? A bit of rust on the wings. A gray stripe at the crown. And so they can play out there in the open. And look at the good times they enjoy gathering coreopsis and partridge pea and beardtongue and goldenrod-seed. Because they didn't get carried away with too many flashy colors. You have to be reasonable.*

To which a separate female elder might have replied, *Even so . . .*

Our pretty, problematic bird, this first Painted Bunting male of the season, fluttered down off the feeder and alit upon a tall blade of grass, more like a reed, tested its rigidity under its modest bunting weight. Wendy and I watched as it skittered up to the top of the reed, which flexed like a bow, as it munched on the seeds bursting from brown, ferny sheaths. I wondered whether it relished those honest-begotten grass-borne seeds more than my store-bought seeds from the plastic cylinder above, whether it enjoyed the flex of the reed more than the stability of the metal perch, enjoyed feeling the impact of its bunting weight in the actual world.

I wondered what species of grass, anyway, grew beneath our feeder to produce that seed bursting from brown, ferny sheaths. Probably millet spilled from the feeder-seed.

When pretty Painted Buntings don't eat my millet seed—from the feeder or from the feeder-seed borne grass beneath—they eat pigweed seeds and bristle grass seeds, and the seeds of wood sorrel and panic grass and spurge and sedge and St. John's wort and pine and wheat and wild rose.

It occurs to me that several of the green birds on the feeder that I assumed to be females along with our sole male might actually have been immature males, that the proportion of females to males that I see out the window each season might not be quite so imbalanced. For immature Painted Buntings, both male and female, sport only the all-green feathers associated with female Painted Buntings. Gradually, gradually, then all of a sudden, red and blue feathers replace the green on male chests and heads. Even their dull green backpacks turn emerald. Which makes me wonder: do immature male Painted Buntings know what's in store for them, that in a matter of months their plumage will undertake a rather dramatic, multi-chromatic transformation? And, if so, how does this bear upon their demeanor with their immature female companions. Do immature males cop attitude as they forage about the pigweed, the bristle grass, the wood sorrel and panic grass. *Out of my way! We may look the same now, but I'm gonna be smokin' hot soon.* Do immature females—and here I presume that they too know what's in store for the boys—cow before the imminent loveliness of their male counterparts, offer a wide berth, or do I have this all wrong? It may be that female Painted Buntings couldn't care less about the ostentatious loveliness of their male counterparts. Perhaps they find all the "peacocking" silly—these outlandish blues and reds and greens. Perhaps it's all they can do to tolerate these puffed up males. They pair up, perhaps, out of sheer pity or desperation. What else are they to do? It's not like they can choose the more down-to-earth Savannah Sparrows or Palm Warblers or House Finches. They're Painted Buntings. What I wonder most generally, I suppose, is whether it creates problems for these pretty, problematic birds, the ocular disparity between the sexes.

After all, it's not like birds are always (or usually) so nice to one another. You don't have to be the keenest of observers to notice that birds do seem to squabble quite a bit for more favorable perches on trees, electric wires, and feeders. Some of the prettiest birds are purportedly among the most aggressive. The dazzling throats and diminutive size of hummingbirds, for example, belie their ferocity. Little assholes, a more experienced birding friend of mine calls them. Further, while most birdsong sounds sweet to our human ears, the truth is that ornithologists don't know precisely what birds mean to say through their vocalizations. As a character in David Foster Wallace's *The Pale King* (2011) cannily observes, "the birds, whose twitters and repeated songs sounded so pretty and affirming of nature and the coming day, might actually, in a code known only to other birds, be the birds each saying 'Get away' or 'This branch is mine!' or 'This tree is mine! I'll kill you! Kill, Kill!'"[4] It's probably important to keep in mind that much of what we see and hear as loveliness in birds is of our own willful, imaginative making. It's a problem.

Something there is, anyway, that can't deny this imaginative work, my hopeful vision that male and female Painted Buntings, immature and adult, interact with one another on mostly amicable terms. I like to think, specifically, that males and females alike enjoy a healthy self-regard untainted by haughtiness, that male birds love themselves for their reds and blues and greens and hold their mates just as dear for their duller green dashed with dark, that female birds love their duller green dashed with dark, too, love their mates just as well, but not more or less well, for their brilliant reds and blues and greens.

I like to think that the birds, anyway, have figured things out.

The evidence mostly suggests that Painted Buntings have done so, after a fashion. When the days grow longer and hotter in their wintering grounds here in south Florida and elsewhere, they light out for more temperate climes northward, seek out brushy roadsides and streamsides, fallow fields and citrus groves, maritime hammock edges and palmetto thickets in places we humans call north Florida, Georgia, and the Carolinas. A separate breeding population favors Texas, Oklahoma, Arkansas, and Louisiana. Bravely, from open perches, the Painted Bunting male sings its "sweet continuous warble," as David Allen Sibley describes bunting song.[5] He courts his mate with great ardor, flashing his bright feathers, bowing and strutting. The female mostly pecks at seeds, unimpressed, or feigning aloofness, but eventually hops toward her suitor to join him in shared purpose. The wings of the male bird quiver with delight. The two set up housekeeping.

Male Painted Buntings don't cultivate harems, evidently. They are mostly monogamous. Like people. I guess.

Both the male and female search out potential nesting sites hidden within dense foliage four or five feet off the ground. Sometimes lower. Sometimes higher. The female gathers material for their twiggy, grassy, leafy, cobwebby, horse-hairy, and rootlety nests: mesquite and elm and osage-orange and greenbriar and oak and Spanish moss. She builds the nest alone. But don't decry the laziness of husbands! He has plenty on his plate. Principally, he defends their turf with great tenacity, showcasing for rival males an exhaustive menu of displays: upright display, bow display, flutter-up display, wing-quiver display, butterfly display. Should such posturing fail, he attacks the intruder, dive-bombing and nipping and pecking. He'll yank out whole feathers between his mighty little bunting mandibles. Again, the behavior of birds, even pretty birds—especially pretty birds—often belies their loveliness. It's a jungle out there. They can't afford to be kind to competitors. Enough, maybe, that they are kind to their mates.

The female lays three or four small bluish eggs, speckled with brown and gray. The eggs hatch in less than two weeks. She deposits all manner of buggy-food into the ever-gaping maws of the chicks: grasshoppers, weevils, beetles, wasps, spiders, snails, caterpillars, and flies. Unlike most birds—less amorous birds—Painted Buntings raise as many as three broods per season. Once the female re-nests, the male will feed their fledged chicks.

Yes, it seems to me that Painted Buntings have pretty much figured out their love business. So brave, these gaudy, lit-up birds, who might have donned drab sparrow colors, but chose a more passionate route. *Life! Life! Life!* Painted Buntings cry with their emerald backpacks, with their flutter-up displays, with their dive-bombing and nipping and pecking, with their cobwebby nests, and (of course) with their sweet continuous warble. John Keats, who would die of tuberculosis at age twenty-five, gleaned in birdsong the indomitable life-force of which he was sorely deprived. "Thou wast not born for death, immortal Bird!" he writes in "Ode to a Nightingale."[6] In reality, most birds—Painted Buntings included—live very short lives. Wallace's more earthbound prose passage on birdsong above (*I'll kill you! Kill, Kill!*) offers a corrective, of sorts, to Keats' unbridled romanticism. Yet I find that my mind seeks out a space somewhere between the romantic and the real when birds flit across my field of view. I'm not willing to give up the notion that these few ounces of feathers and bone and flesh, as Keats' ode suggests, epitomize life lived full-bore. *Life! Life! Life!* The problem with pretty birds is that they constantly put me to shame with their bravery, their unwavering self-assuredness, their moxie, their lives lived full-bore. Self-doubt doesn't seem to be such a big thing with them. "We're never single-minded, unperplexed, like migratory birds," the poet Rainer Maria Rilke writes.[7]

I'd like to believe that my perfectly ordinary suburban life is a brave life, too, yet most social indicators tell me otherwise. A timid life, my home-centered life, from a certain vantage. We are still a young country that celebrates new beginnings, new journeys, constant reinventions of the self. "A vagabond wind has been blowing here for a long while," Scott Russell Sanders declares in *Staying Put: Making a Home in a Restless World* (1993), "and it grows stronger by the hour."[8] I've watched several close friends and colleagues set sail upon this wind for new jobs, new cities, new spouses, observed these departures with sadness and, admittedly, a tinge of envy and un-birdlike self-doubt. I'm not completely immune to the spirit of our times. It may be why I'm so antsy and irascible with my family, sometimes. I'm not single-minded, unperplexed. It's a problem. Like Sanders, though, I've mostly ducked the vagabond wind, hunkered down for nearly twenty years in the same city, at the same embattled state university, with the same spouse. From whence did such stick-to-it-ness arise?

The example of birds has probably reinforced my home-centered inclinations, and from a very young age. My parents, like many American parents, moved my siblings and me around quite a bit, forsaking the east coast for the west coast when I was five. I was just old enough during this uprooting to have felt that it was an uprooting, to have felt that home was something we had left behind, something to recover and hold dear. This may be why I marveled, growing up in Southern California, at the story told by several of my elementary school teachers of the Cliff Swallows' yearly return to the Mission of San Juan Capistrano. Each year on the same exact day, St. Joseph's Day, March 19th, townspeople would raise their eyes to the sky to welcome back the Cliff Swallow flocks from their winter home in Argentina. The swallows would immediately set up housekeeping, constructing their tiny mud nests against the crusty ruins of the Great Stone Church. The swallows knew something essential, something strong enough to hold year after year after year, and on the exact same day(!), something about the power of their home-place.

While I never convinced my parents to take me to the Mission San Juan Capistrano on St. Joseph's Day to observe the swallows' homecoming, I did somehow (and to my continued amazement) persuade them by seventh grade to let me raise homing pigeons in a backyard coop. I constructed the coop (shoddily) with a few of my pals over several weeks before buying my first two pairs of homing pigeons from a local breeder. The man warned that I couldn't release the birds from my coop for their first flight until after they nested and their first chicks had hatched. If I released them before this time, they would simply fly back to his coop at the other end of the San Fernando Valley. It was an interesting fact about homing pigeons, which remains interesting. Home is where they raise their young. I heeded the fellow's admonition and kept the birds in the coop until their first chicks fledged.

I still remember the thrill of that first release. Riding our knobby-tired bicycles, my friends and I transported the bird families in brown grocery bags up into the chaparral hills several miles from our neighborhood. (In the years since, these hills have mostly been paved over and developed, planted with stucco and Spanish tile homes, but back then it was wild land, or near-wild land, teeming with rattlesnakes, coyotes, and Red-tailed Hawks.) We released the birds and sped back home, the sunbaked trails giving way to a long downhill stretch of asphalt road. The homing pigeons, miraculously, were waiting for us in the coop already, their chests puffed up, strutting and cooing. They knew what the swallows knew. Home is where you raise your young.

I don't raise homing pigeons anymore. Instead, I keep watch for the reliable winter return of songbirds, especially our Painted Bunting flock. Despite manifold competing factors that vie for my attention—successfully, more often than not, I'm afraid—I still look toward the buntings, these creatures who I know couldn't care a fig about me, or us, but continue to offer essential instruction, abandoning their twiggy, grassy, leafy, cobwebby, horse-hairy, and rootlety nests at roughly the same time each year to make our home their winter home.

The five female buntings this recent, harried morning skittered off the feeder, eventually, swept up the male below and disappeared into the scruffy firebush and necklace pod foliage. "Guess they're finished," my wife said, then sighed. A curious look painted her face. A furrow between her eyes betrayed mild disorientation. The furrow asked, *What was it we were talking about? What unpleasant domestic business must we finish up this morning?* I remembered our business together, our little spat over who would watch Eva this afternoon. I'd like to say that the glimpse of our first Painted Bunting male of the season had put me in my place, immediately, that I stepped up to the rather modest demands of parenthood with alacrity. But I can't quite make this claim. I conceded the argument, but I could have been a whole lot nicer about it. Only now do I wonder whether my insistence on nursing a foul mood over something so small amounted in the end to the vanity I foolishly project sometimes onto pretty birds, whether Merlins of my own making loom over all that I hold dear. Only now do I glance back, in my mind's eye, from the glass door toward the remnants of breakfast on the table, toward our family room, toys strewn all about the rug like leaf litter, toward our small daughter looming over her puzzle—our panic grass and wood sorrel, our cobwebby nest, our fledgling with her ever-gaping maw.

Chapter 6

Red-headed Love Child

J. Drew Lanham

"Daddy! Daddy! Daddy!" Someone was trying to get my attention, but as far as I knew my legitimate spawn were more than a thousand miles away in South Carolina. Here I was, in the middle of God-knows-where, Nebraska, and standing not ten feet away was a little blue-eyed boy with curly red hair claiming me as something I wasn't. I froze in the midst of a difficult decision, choosing between Wrigley's Big Red and Juicy Fruit chewing gum. Now there was a different dilemma beyond spicy-sweet or fruity deliciousness. For a moment I was cornered by a three-foot-tall child whose blue eyes could've been left behind by the glaciers that once covered this part of the world. Maybe three or four years old at most, the cherub stood there, confident of the candy aisle proclamation that I was the one so designated to pick him up, squeeze him tight, and twirl him around in some sort of joyous father-son reunion.

Of course, I had no intention of getting anywhere near this toddler. A big black man picking up a Caucasian kid in what seemed the ivory center of the lily-white Corn Husker State would've likely resulted in sirens wailing and the deployment of local, state, and federal authorities—or so I thought. But in spite of the instinct telling me to put as much distance between the red-haired boy and myself as quickly as possible, I stood there, staring back; my brown eyes met his blue ones. There was something about him that was, well, different. Maybe it was the red 'fro. Maybe it was his not-so-white skin.

Soon, a rather attractive auburn-haired white woman found her child in the middle of making a connection with a man of color. I suspect it was not a first for either of them. She grabbed the toddler. "He's not your Daddy! I—I'm sorry sir." With that and a few more hastily constructed words of apology, embarrassment, and anger thrown in for added effect, they both went on to buy whatever convenient thing they were in a convenience store to buy. I chose Big Red. At that moment, spice seemed the way to go.

Back in the rental car that had pushed across a good deal of central Nebraska, I retold the story to my travel buddies, Matt and Steve. They found the event that had occurred on a gasoline fill-up at first unbelievable and then unbelievably funny. The happening provided some comic relief as the miles wore on. I speculated about what trashy television show I'd end up on in twenty years with some burly biracial Cornhusker man claiming a life less fulfilled without me. Steve and Matt promised that "what happened in Nebraska would stay in Nebraska!"

It wasn't my first birding trip out west, but it might become my most memorable if random children kept stepping forward to claim me as their father. The curly coif and the not-so-pale skin pointed to the boy's other half being of a darker hue. Given that I'd only seen one black person since I'd been in Nebraska, I began to recalculate my thumbnail homogenous analysis of the state.

Ask 100 people anything about Nebraska, and you'll get "corn" somewhere in the first three words of their response. If you'd asked me about Nebraska just a few weeks before my first trip there, my answer may have been more complex and included something about prairie birds, the famed college football team in Lincoln—and corn. And so when I landed in Grand Island, Nebraska, a place probably not otherwise noteworthy for much more than being smack in the middle of Nebraska corn country, I knew that we were there for more than grain. We'd journeyed out in the last days of March to witness the tail end of one of the great ornithological wonders in North America: the northward migration of Sandhill Cranes along the Platte River.

For probably 10,000 years or more, the tall, steel-gray birds have thrown their unmusically beautiful calls across the shallow floodplain that is now in the heart of America's corn and burger-producing breadbasket. I'd seen Sandhill Cranes before, but in Grand Island, they did not seem to occur singly anywhere. As we scouted on that first day for cranes, it became apparent that even at the waning end of the migration *rare* was not an accurate descriptor for the undulating ribbons of birds that cruised the skies, circled on rising thermals, and dropped like paratroopers into stubble-riddled cornfields lying brown and rich with river-run soil. In the air they were gracefully buoyant and powerful fliers. On the ground they were just as stately—walking, stalking, dancing, and prancing as crane-kind does. When you are surrounded by cranes it is easy to understand how the family of birds have generated awe and worship around the world. As the day closed we watched the evening roosting ritual as a half dozen here, twenty or more there, flew out of the setting sun like legions of phoenix to roost on the sandbars in the shallows of the Platte. By dusky dark there were thousands of long-necked, long-legged silhouettes gathered together and calling in the most pleasingly discordant chorus. It was a sound etched deep in my soul.

I recalled the crane music, the soft rush of the river's flow, and the calling of frogs as I fell asleep that night.

We wrapped up our crane filming job the next day and headed north towards South Dakota for a date with Greater Prairie-Chickens. As we sped along highways, my initial impression of Nebraska catering to corn was borne out. The landscape was like rough five-o'clock shadow with the stubble evidence of the combine's shaving the previous season. Soon the next season's planting would go in, and green fingers of the grass that would become agricultural gold would spring up in the no-plow rows. But for now, I was tiring of the sameness. Cranes were beginning to become an afterthought when the miles mercifully scrunched the foreverness of flat fields into rolling hills covered with grass and occasional stands of pine forest. In between the rises, there were crevices and draws that pulled water into small ponds and larger lakes that were dotted with ducks. Big cottonwood trees stood here and there, testament to wetness lying not so far beneath the sand.

We'd entered the Sandhills; ancient high dunes still bore the evidence of wagon travel in the visible parallel ruts that stretched across the arid expanses. Every now and again, some small town would pop up. With a sign declaring populations in the tens or maybe hundreds at most, the apparent requisites for civilization included a grain elevator, a church, and maybe a gas station. The town where my biracial would-be love child and I were reunited was a metropolis compared to anything else we saw out there. Outside of the towns lay a landscape that felt wild and sometimes foreboding. The dunes went on and on. Other than the barbed-wire fences festooned with old tires, there was little else indicative of human presence.

We saw American White Pelicans soaring like pale pterodactyls over the duck-dotted potholes and counted Swainson's Hawks like mile markers sitting tall on every other telephone pole. Nebraska sits on top of the Ogallala Aquifer, an enormous underground lake that slakes the thirst of much of the Midwest and its never-ending (and unsustainable) desire for water to keep the breadbasket productive. There were birds everywhere, but our target wouldn't be so easy to spot from roadside voyeuring. We'd have to rise early and seduce the presence of our targets in silence.

Prairie-chickens, in one form or another, ranged across almost two-thirds of the eastern and midwestern U.S. in the not too-distant past. With various regions possessing their own uniquely adapted species, "chicken" numbers have declined dramatically as prairie habitats have been plowed under or overgrazed.

We were lucky enough to be in a place where the birds could still be found with patience and a bit of luck. What we wanted was more than a look, though. We were there to see Greater Prairie-Chickens dance in a millennia-old ritual that made sure more prairie-chickens would be conceived and hatched to keep the prairie pulse pumping. To see the dance required a predawn hike across the sandy terrain and whispered silence in a blind built by the federal wildlife folks for the purpose of chicken-spying.

As much as birders depend on their eyes, many of us would be blind without our ears. The morning of our rendezvous, the ceaseless prairie winds vibrated with northern leopard frogs snoring and chuckling from the draws, pheasant cocks crowing, and American Bitterns pumping as the stars dimmed in the coming dawn. It was more dream than real. Only a few days before, I'd been in the spring-greening forests of the South Carolina piedmont and now I was in a world so different that it seemed otherworldly. Here, the grasses were the old growth. Underfoot and all around me, life abounded in the rolling hills and wetlands in a diversity that was almost uncountable. Maybe it was just the wind that drew the tears from my eyes. Maybe it was the privilege of the place that made me choke back the joyful weeping. Though I'd never set foot in this place, I felt a part of it—rooted deep like the grasses in the sand.

Being in the midst of it all was like having a Muse sing the sweetest lullaby. Once in the blind, I closed my eyes—not to sleep, but to somehow soak it all in: the sounds, smells, and sights. I prayed to some force beyond me in thanks for all I had seen and the chickens I could not see. And then *wooooooooom*—from somewhere out there a low moan floated in on the wind. It came again, *woooooooooooooom wooooooooom;* gentle and ghastly—*woooooooooom woooooooooom*—prairie-chickens!

Our attention focused on the lek, a sandy dance floor the chickens had cleared and displayed on for generations. With the blind not more than a few feet from it, we had a front-row seat. The "wooming" had stopped when something suddenly appeared center stage! Squinting hard to somehow squeeze light onto the subject, I made out the form of a plump fowl that, yes, looked chicken-like. It was my first Greater Prairie-Chicken.

There was nothing particularly spectacular about it. In size and character, it did look like a small brown-speckled hen. The bird pecked around a bit and was soon joined by another. Once the second chicken flew in, any resemblance to what farm folks back home call "yard birds" disappeared. The two birds faced off on opposite ends of the oval dancing ground, plumage puffed out with two tufts of feathers standing out like horns on their heads. They fanned out their tails like miniature strutting turkey-gobblers. Two tangerine-colored sacs swelled up like small balloons on the sides of their necks. Their oversized eyebrows flamed yellow in the freshly risen sun. The puffing and fanning soon became a dance; a twirling, foot-stamping, wing-dragging dance daring the other bird to do something more.

As daylight finally broke, a crowd had gathered around the stage. We weren't the only watchers. A half dozen hens and a few other cocks had sneaked in to see the show. The peekaboo was serious business for the chickens—a matter of life and death. The ladies would ultimately choose a dancer to mate with and produce the next generation of *Tympanuchus cupido*—little drumming lovers. The usually secretive males took few risks in combat, since bluffs were the way evolution made sure the species didn't cut off its nose to spite its face. The greater risk came in the exposure to predators that were probably also watching. Bout after bout of display and occasional contact continued for hours, and then, just as suddenly as it began, it ended. The birds left as they'd entered.

We had our show on film, and I had the thrill of a birding lifetime. Pheasants crowed, frogs chuckled, and the bittern boomed beyond the dawn clock it usually punched.

Birds connect me to strange places in ways that make me feel at home. Try as I might to explain it to those who've never lost the hours watching feathered life go about its business in some coastal bay marsh or mountain cove forest, it's a hard point to press home. Home is indeed where the heart is. For me, that means that I can find comfort in piedmont old fields and new sunrises on prairie sweeps. Southwestern deserts and south Texas palm groves feel more familiar than not.

I consider myself a migratory creature, traveling back and forth across the country to find birds and the wild places they live. Places like the rolling Nebraska Sandhills, the sky islands of southeast Arizona, the cold running streams of Yosemite, the flat-topped buttes of Montana, the Texas Big Bend, and the Kansas tallgrass comprise the many places "out west" where I've been to see birds. They are like many of the places I go to see birds anywhere in the world: far off the beaten path and mostly populated by white people. In my crisscrossing and wild wandering I seldom find other black birders. In some ways I used to see myself as a pioneer of sorts, a first among the throngs of seekers in far-flung places wandering underneath big skies or witnessing windswept vistas.

After our chicken and crane odyssey Matt, Steve, and I took the long way back to our airport in Grand Island. We found an out-of the-way wildlife refuge called Fort Niobrara. We drove through a loose herd of bison with Afro-headed bulls standing taller than the car. In the visitor's center I found a revelation that dispelled any notions I had about being the first black man to venture into northern Nebraska. Along with a giant stuffed bison that made me feel tiny, there was a historical display with grainy depictions of soldiers. I looked closer, thinking at first the old photograph was simply casting shadows on the faces of the regiment. But then the words brought the truth home. These were black men! Sent to the far ends of the late nineteenth- and early twentieth-century western frontier to hold the line and press the manifest destiny of a growing nation, black American military men, the famed Buffalo Soldiers, had been here long before I was anyone's thought.

More than 100 years ago, black men of the U.S. Army's Ninth and Tenth Cavalry and Twenty-Fourth and Twenty-Fifth Infantry followed orders and endured the extremes of heat, cold, dust, mud, insects, and disease that often plague the out-of-the-way places I go by choice to find birds. In between the daily tasks of surviving rampant racism from the U.S. Army, skirmishes with American Indians fighting (rightfully) to hold onto homelands, and incursions from Mexican patriots (trying to understandably reclaim lost homeland), I'm sure there wasn't much time for the leisure of watching birds or rising at dawn to witness a prairie ritual. But then again, this Nebraska trip was breaking brain barriers I'd long held as dogma. Maybe I was giving these brave men short shrift.

I'd like to think that all of us, regardless of circumstance, find some way to appreciate the wonders of the world around us. Maybe on an evening watch, a blue-coated black looked skyward during his to-and-from march, his ears catching the trumpeting calls of uncountable numbers of Sandhill Cranes coming in to rest on the sandbars of the shallow North Platte. Maybe the trumpeting chorus makes low pay and lack of respect fade for just a moment as the sun glances off broad gray wings. Perhaps a brown-faced horseman, the son of a slave now free, wears knee boots dusty from the trail and a broad-brimmed hat slouched low over tired eyes. He sits tired but tall in his McClellan saddle after a patrol through the Sandhills. As the troop pauses to let tired mounts rest and drink from a pothole pond, he ponders the spectacle of a snipe climbing high and twittering to the heavens one minute and plummeting to the ground the next. Maybe there were a few opportunities for duck dinners to break the monotony of beans and bread. A two-lined phalanx of free black men on bay mounts must have been impressive, winding their way across rolling grassland, the eerie cries of Long-billed Curlews and the sweet songs of Western Meadowlarks pacing creaking saddle leather and jingling bridle rings. A bold pair of acrobatic Scissor-tailed Flycatchers takes advantage of the bounty of bugs kicked up by the horses' hooves, and a trooper marvels at the boldness and grace of the long-tailed birds.

Black people, or people of any color, surely could not have ignored the beauty of towering sequoias or the breathtaking beauty of sunrises and sunsets painting the Rocky Mountains and Great Plains. No, I don't have any café au lait love children in the Sandhills of Nebraska or anywhere else for that matter. The blue-eyed boy was mistaken in that matter. But then, maybe there's more to what the little one saw than I give him credit for. Perhaps somewhere in the colorful memory of that curly-haired boy, there's a Buffalo Soldier who fathered a memory we all need to claim.

Chapter 7

Crane, Water, Change

A Migratory Essay

Christine Byl

Neighbors

Near the end of August, I wait for the first sound—a few cranes high up, barely visible. Not flocks, not yet. In Interior Alaska, through early September, it gradually builds—the number, the sound—until a crescendo in the third week, a raucous noise that vaults past beautiful. Sandhill Cranes on the move are startling and epic. A reminder that wildness is not always silent.

By early October, numbers thin, lessening until, again, a few birds high up, barely visible: migration's parentheses. The last cranes exit closely followed by the first snow. "Hurry, guys!" I cheer them on as Arctic storms barrel south from their nesting grounds. "It's getting cold. *Go, go, go*!"

The first time I saw sandhills en masse, I lived further south, on Prince William Sound. Notable flocks flowed overhead in spring, but they intermingled with every other noisy bird that passes through the Copper River Coastal Plain, and it was hard to focus on cranes alone. Now, I live outside the town of Healy, abutting Denali National Park's northern border, and some mornings cranes shade the skylight above my bed, the first living things I see when I open my eyes. In our arid, empty sky, cranes are the stars.

My neighborhood is knit by sounds: mosquitoes, engines, birdsong, wind. We humans seek quiet privacy on our own two acres, or twenty, but at the edge of taiga forest, stunted black spruce and low tundra plants do not muffle sound like denser forest does, and we live amidst each other's daily noise. Howls from close north mean Jared hooking up the dog team. A nail gun early is Jess at her rafters, hoping to be weathered in before snow. Neighbor kids shout and fire BB guns too close. Approaching tires on the gravel road say *truck* or *station wagon*, *local commute* or *tourist drive-by*, *trailer* or *dump truck* or *ATV*. We add our voices—bikes stopped in the road, car doors flung open to talk—waves, short honks, or a finger lifted from the wheel, the quietest local hello.

This neighborhood is a rural subdivision, which makes it a home for us. Pocketed with wetlands, it's also a home for birds. Up the road lies Eightmile Lake, bowled in muskeg. In springtime, Panguingue Creek tributaries flow loud enough to hear over a chainsaw's drone or drifting country radio.

Resident birds offer up their noise with ours all year long: a Great Horned Owl pair hooting in the aspen grove, loudest in May, and the magpies, tricky ventriloquists, hollering in every voice but human. Ravens dip into the dog's bowl and crow about the soggy kibbles floating there. In spring, migratory birds join the cacophony. Robins drink from puddles and sing bright orange songs, novel after winter's limited palette hush. Dark-eyed Juncos *chip* from black spruce limbs. And cranes pour overhead in pulsing Vs. "Welcome back!" I shout from the porch.

As busy summer closes into fall, the humans, not prone to standstill, finally pause to catch up. An annual harvest party has sprung up in our neighborhood; the first weekend in September, we gather for a local food potluck. Makeshift tables of sawhorses and 2 x 10s groan under regional weight—moose backstrap and flounder ceviche, smoked salmon three ways, zucchini noodles, sauerkraut, sourdough bread and beer (some home-brew, some PBR). We debrief: trips taken or aborted, too much work, too little, a good garden harvest, a bad fishing year, the summer's biggest tomato.

Often, it rains. Bonfire, giant tarp, layers zipped up to our chins. One year, right before dark, the rain lifted enough to allow the day's clearest shaft of unfiltered sun. Cranes streamed from the west and filled the lit sky with chorused uproar. The party went still. Together, we watched: Susan and David who have lived here for twenty-eight years; the romping pack of kids; someone's parents visiting from Wisconsin—all our faces uplifted. I can't remember another time outside with thirty quiet people. The growing dusk was so loud with cranes we couldn't have heard each other speak anyway. I watched my clustered neighbors, chins tipped upward as if for a blessing, though the cranes were too intent on travel to bother with dispensing grace, and I wondered, is everyone thinking what I am, under the sky that spans our home? This is what I'm thinking: of course, the cranes are here. They are our neighbors, too.

Three Things I've Learned About Cranes

One: The birds that migrate over Healy are Lesser Sandhill Cranes, of fifteen crane species worldwide, the only one that breeds in Alaska.[1] These Central Flyway sandhills (also called the Midcontinent flock) head southeast in autumn from the Yukon-Kuskokwim Delta, Eastern Siberia, Norton Sound, in five to six distinct streams of migration converging just east of Healy. The Alaska Range forces the birds northeast over the Upper Tanana Valley before they head south again, wending through Canada, east of the Rockies, over Platte River country to winter grounds in the Southwest and Mexico.

In the mid-'90s I lived in Western Montana and I remember seeing only a few cranes, stragglers off course from their migration path slightly further east. The average lifespan of a crane is twenty to thirty years (the oldest known, thirty-six years and seven months),[2] so it's likely that some of the birds I saw in Montana are part of the flock that migrates over Healy now. That a single bird could connect two homes, more than 2,500 miles and nearly two decades apart, blows my mind. Me in my twenties, new to the West, and me in my forties, at home in the North. Different worlds. *The same bird.*

Two: Cranes, like most creatures, have multiple calls. A contact call states *I am here*, the adult's a soft purring growl, in chicks (or colts) a tiny *peep*. A distress call is strident, to draw attention, and a guard call intensifies warning—a single high blurt, a crane alarm. (A colt only a few weeks old can give a guard call; even young can recognize threat.) Sandhills' unison song is one of the most distinct bird sounds; the female sings two notes, the male one, and their entwining call-and-response duet reinforces bonds and stakes territorial claim. Kim Heacox writes about these monogamous birds, "if loons invented the music of being alone, cranes invented the music of being together."[3]

To describe a crane's call—any one of them—pushes the limits of language. The sound is nearly impossible to describe. *A thumbnail dragged along a taut piano string, frictioned stops and starts*. All metaphors fall short. *The hum of a small plane's engine beyond the next ridge*. I sit at my desk trying—*an old-fashioned wooden ratchet noise-maker at a soccer match*—but I can't make you hear it—*children shouting on the playground, one odd voice above the rest, creak of an old see-saw in the background*—and then my mouth and throat open and I croak and warble instead. That which cannot be described by the mind tries to lodge itself in the body. Sound is not a sentence you quietly read. It requires a living world, and presence. Of cranes, making, of me, hearing. *Listen*.

Three: The Athabascan word for crane is *dildoola*, lilting the onomatopoeia of their song, and the word for cranberry is *dildoola baba*, meaning "crane's food."[4] English echoes this trajectory, our word from Low Germanic origin, *kranebeere*, named because the plant's flower, petal and stem resemble the head, bill and neck of a crane. One word connects eater to eaten, a berry connects Athabascan and European, a bird connects berry to mother tongue. Cranberry and crane's call are woven in with boggy sphagnum, the wetland that supports them both, where one grows and the other feeds. This is my autumn stew, taste and sound basted by wetness and coming cold. The bustle of motion—*fly, pick*—and the sure advance of winter's stillness.

Wet, Dry, Ice, Mud

One cannot write about Sandhill Cranes without writing about wetlands. Summer nesting, breeding, hatching and foraging in the far north;

migratory rest stops; wintering grounds to the south—cranes can't *be* without wetlands to be in.

Arctic wetlands are a crane buffet. Voracious omnivores, cranes surface feed and also use their bills to probe and plunge and strike and nibble, feeding on what's near. Cranes are opportunivores: they'll hunt-and-peck for lichen, mosquitos, even a vole if they can catch one. Cloudberries, dwarf birch, cotton grass, reindeer moss, Labrador tea. Crane food grows in saturated tundra marshes and along riparian zones: the wet edges.

Arctic wetlands are nurseries. Sandhills often breed in open tundra, nests perched atop high spots: sandy knolls, raised mounds, or dry islands on ponds. These areas shed snow first, and high ground ensures good visibility. When a mating pair builds its nest, the male shops for grasses and sticks, tossing them to his mate, and she builds materials into a structure. Cranes use the same nests over years, raising generations of young from a familiar swampy home.

Wetlands are rest stops on the Central Flyway turnpike—flocks hone in on braided river bars from thousands of feet above, often beckoned to the ground by the presence of other birds. And wetlands are community centers, where cranes cruise for mates. They preen and high-step and duet at the sedgy edges of things, their favorite swamping grounds.

The sub-Arctic is a dry climate, but not a dry place. Sans snowfall, the north side of Denali would classify as desert, receiving less rain per year than Salt Lake City. Despite such aridity, the surface of the region is intermittently saturated, thanks only to permafrost (ground that remains frozen for two consecutive years). Such cold dirt takes many forms: continuous, discontinuous, thawing, sporadic and static permafrost. The balance of freeze and thaw creates trapped surface water that can't percolate as it would in unfrozen soils. Think of the thawing layer of permafrost as a water tap and the frozen layer as the bottom of a bathtub. These shallow places evaporate more quickly than deeper bodies of water, so cranes' choice feeding grounds are particularly vulnerable. The amount of ice in permafrost and the amount of permafrost in soil affect the speed of thawing, but generally speaking, warming average annual temperatures mean a pulled plug and a swirling drain in currently productive wetlands.

The permafrost front—a band across Interior Alaska where permafrost is consistently degrading—extends laterally from Kobuk to Bettles to Tok, roughly perpendicular to the cranes' migratory path. Like treeline, the front is affected by elevation and latitude, and it's migrating northward as the climate warms. In a time of flux, wetland habitat becomes a moving target. Cranes will lose existing range in some places, and gain habitat in others currently too frozen to be productive. In Denali, dry-up is already clear—ponds have receded, lakeshores contract. Add to thawing permafrost the fact that the arid summer's evaporation rate will likely exceed precipitation, and a net loss of surface water results. Exact gains or losses are hard to predict, as is true for most hypotheticals in climate modeling, but Arctic warming itself is not a projection. It's happening, faster here than anywhere else on earth.[5]

The factors affecting crane habitat are complex and, as with all natural processes, interwoven. Interior Alaska's wetland recharge happens via spring snowmelt. From April to June, Denali's harsh winter temperatures ease, and ice and snow begin to deteriorate. This distinctly Alaskan season is known as "break-up," and the best scenario for wetlands is the "catastrophic" variety, where snow melts fast, river ice jams give way, and heavy rains fall. The still-frozen ground can't absorb it all at once—the tap is on full blast, and the tub fills quickly. A protracted, slushy thaw, on the other hand, is a trickling faucet, and standing water percolates faster. The tub never fills. In recent years, break-up has trended toward the latter, a gradual shrug into spring called "mush out" instead. Less water on the ground in spring hugely affects migrators, who pass through during that exact window.

This century has seen multiple small warming periods, couched in one long warming period (the Holocene Age), but the current rate of change has accelerated far beyond the background rate against which species have evolved. This past March was the warmest on record in Interior Alaska. Conservative predictions point to a 7–8° F temperature shift in the Arctic. If you are a crane, this is a frightening forecast.

Audubon currently categorizes Sandhill Cranes as stable but climate-threatened and their 2014 Climate Report is sobering. "Of the 588 species Audubon studied, 314 are likely to find themselves in dire straights by 2080."[6] (In North America, nine bird species have gone extinct in all of modern times.) The Climate Report predicts that cranes will remain viable, but not without major concerns. Melanie Smith, an ecologist and cartographer at Audubon, lists questions researchers ask when trying to gauge species viability. "How adaptable will they be? Do we think they will stay in one place and suffer? Or will they find new habitat?"[7] The Arctic will likely retain sufficient crane habitat, though its location may change; pressures in the Lower 48 are more intense, including climate factors as well as human encroachment.[8]

Climate scientists concur on the general trajectory of warming and its effects but as Smith puts it, "the larger scientific community does not right now have a sure way to predict" which species will thrive and which will collapse. Projections are at best educated guesses, proven or disproven only as the future becomes the present. To Smith, this is the conservation's broad challenge, in light of accelerated, unprecedented change: "We have to do something with the knowledge we have. We can't get bogged down by what we don't know or can't predict." Certainty may come too late for many birds. "Things haven't changed this fast before," she says.[9]

To mitigate so many unknowns, Smith notes, we must protect the places we know are important today even while identifying places we think will be important tomorrow. If birds are limited to current (waning) habitat, the pace of change won't leave space for adaptation. Large tracts of land could help absorb potential changes in birds' behaviors, which is why Audubon focuses on land conservation. Protected acres across a diversity of habitats may buy birds time and space to adapt.

Enter the "Portfolio Effect." Just as financial planners advise that risk is best managed by investment diversification, scientists defend nature's broad portfolio: biodiversity. More species, more habitat, more interactions between healthy animal populations, all mitigate risks like drought, flood, population crashes and weather extremes. To buffer climate pressures on birds, Audubon buys "stock" that is doing well now (currently productive habitat), and some whose value will rise (projected future habitat.)

Cranes would be wise to hedge their bets; they recover slowly from population lows. Some bird species lay many eggs where only a small percentage survive, but cranes birth one or two chicks per year. Like human infants, each colt is prioritized; there is no planned expendability. When a population is threatened, this can be a weakness. But cranes also have major strengths. Their migration patterns seem to be learned, not genetically hardwired, so they can incrementally change behavior over time, each leap a stepping-stone to survival. And cranes have species longevity, a genetic windfall borne of riding out eons. They are the longest surviving species of any still-existing bird, a winged dinosaur whose fossils date back 2.5–10 million years. Nearly hunted to extinction in the 1930s, sandhills, once protected, rebounded to their current relative stability. They have been surviving—with humans, without them, in cold temperatures and warm—for ages. Cranes are ancient and versatile and accustomed to change. They offer lessons for living amidst flux. Protect space. Diversify. Adapt.

The contemporary Earth is huge and far-flung, but it is also tightly bound. Water ties distant places together, a fact we learn in grade school (water, that vascular system of the world) and revisit as adults ("we all live downstream"). Precipitation in one bioregion becomes run-off elsewhere; one latitude's ocean currents influence another's snowfall.

Cranes connect ecosystems, too. The air and water that flows between us is mostly invisible and aggregate, but cranes are distinct, trackable, and audible. They come to our place from someone else's place, and they return, announcing their arrival and departure. They affect and are affected by regions we never see.

In addition to the biological wonder and the aesthetic spectacle of those flocks, they also do metaphoric work. Cranes are ambassadors for otherness. They trigger us to imagine beyond our borders, and to beckon the far-off closer in. Oregon poet William Stafford wrote about cranes:

> They reach for the land; they stalk
> the plowed fields, not letting us near,
> not quite our own, not quite the world's.
> People go by and pull over to watch. They
> peer and point and wonder. It is because
> these travelers, these far wanderers,
> plane down and yearn in a reaching
> flight. They extend our life,
> piercing through space to reappear
> quietly, undeniably, where we are.[10]

Four Questions for Cranes

One: How do you learn to dance? To jump and lurch and twist, a feathered yogi? You combine awkward and graceful like a teenaged athlete or a young moose. Science hypothesizes that your dance is both hardwired and learned, the way some birds find song. A finch's tune improves as it grows, as do your steps, gangly and stumbling at 20 days old and weeks later, markedly improved, half skimming rhythm, half cavort. You dance to communicate, to attract a mate, and sometimes you seem to move for sheer pleasure—that meditative stalk, neck dipped and wagging, wings out like the arms of a balancing child. When threatened—eagle above, fox on the ground—you turn acrobatics to weapon. Kick, flap, leap to fend off menace. I understand, how the body is differently moved by love or anger.

Two: You are particular to specifics. You choose certain wetlands for breeding, raising years of colts from the same nest. You graze favorite fields each winter, gleaning harvested corn stalks and fallen threshed grains. Your migration patterns are so set in stone that for decades, you will stop to drink at the same river bar. But your travel patterns are elective, not rigid, like some other birds (the godwit, that famed long-haul flier, seems to be genetically encoded). So how do you choose—when to leave, which pond, field or hummock for a nest—the first time? When you return to a familiar place, does it feel as it does to me, the respite of home after away?

Three: Do you get thirsty when you travel? Hungry before it's time to stop? Discouraged? Do you have to ever force yourself on in the face of headwinds and lowering clouds? Once, traveling on a glacier in pelting snow, a week into a ski trip, short on calories and rest, I thought of you. *Imagine you are a sandhill migrating.* I angled my torso into the wind. I would love to know if your endurance is always innate, or sometimes, like mine, a decision. Do you feel relief when the flock arrives on winter ground, where you can eat and stay and rest and eat and rest?

Four: Do you know how singular the sound your throat makes? Your brain is conditioned to begin with the right noises (not those of a loon), and then you build your vocabulary imitating older birds. You practice, learning to sing while you learn to dance. Your calls get clearer and stronger, like your steps.

You are among the loudest birds in the world, that resonant tenor from the shape of a coiled-up, bony-ringed trachea in your sternum, a hollow snake, which, if stretched, would reach from your beak to your feet. You sing while you mate and while you fly, while you feed and even while you rest—Hank Lentfer tells me, "cranes are talking all the time on the nest." I wait the year for the harmonic sound of you in flight. Your sonorous rattle, that woodpeckerish rusty hinge. Do you hear your own voice, or the merged songs of your flock? If birds went extinct—any, or all—how many sounds would be lost, like dead languages whose native speakers no longer exist to mouth them? Extinction must bring a quiet I hope I never hear.

Neighbors, Again

Animals know how to use what the world offers. I love how suited creatures are to their habitats, to a level of survival that includes a core of *thrive*. I admire a wolf's coat—water-resistant and warm, comfortable in temperatures from 80 above to 40 below—while I change in and out of ten layers from season to season. I notice how a wood frog's spotted skin conceals it among dead aspen leaves floating at the edge of the pond where it breeds, how its blood enables it to freeze solid in winter and thaw only when it is again safe to be cold-blooded and thin-skinned. I covet a Snowy Owl's furry feet. I marvel that a bear sow's body delays egg implantation until it has enough fat stores to grow a fetus.

When I watched the cranes fly southeast last fall, my unspoken admiration took a further turn. A fully-worded question sprang to my mind as I looked at the sky, and it recurred to me every time I heard their racket: *What is it like to find the world sufficient?* I've asked it over and over in the months since, a koan I worry at, hoping for the sharp stab of insight.

Cranes expertly adapt themselves to what the world offers. They plan their travel routes around water, patching a migratory path across linked wetlands. Unpowered fliers (birds that stay in the air with an up-stroking wingbeat, not the more common and efficient down-stroke), they compensate for their slower pace by leveraging wind and the heat that rises off thermal masses—water, open tundra bogs—to help lift them airborne with minimal flapping. To conserve energy and maximize loft, cranes often leave the ground at the warmest part of the day, spiraling to rise wherever air currents allow. Even the way cranes site their nests is passive design at its best. Atop elevated humps, water drains to the base, leaving a high-and-dry spot for a nest, moisture and food within reach. This is how you find the world sufficient. You make a bed out of whatever soft thing is near. When there is water, you drink it. When it is dark, you rest, unless dark helps you, and then, like an owl or a bat, you prowl.

One of the greatest tragedies of climate change, one we can never explain to animals, is that we are shaping a world that will no longer provide for them. The simplest cause of extinction is that the world becomes not enough, or, too much. We have recently begun to admit that we are vulnerable to the same fate. Like most species, humans will not be around forever. Our ingenuity and plastics and desalinated water and space travel occasionally convince us that we do not evolve or exist within limits, but if the world becomes insufficient, we, too, will diminish and die out.

I am unused to true scarcity, the kind that triggers thoughts of extinction. My life, like many people's I know, flirts with self-sufficiency—eating by the seasons' dictates, living off the grid, harvesting wood and sun for fuel—and still, I rarely confront need. A lean berry year is softened by remainders of last year's epic bounty still frozen in Ziplocs. A cloudy month means we buy more gas for the generator. When one summer brings not enough salmon, we trade with friends, firewood for fillets, and we eat less fish, more of something else. For most of us, scarcity means there are no ripe avocados at the grocery store. Perhaps only when we have to confront the stark reality of not-enough—the hunger or thirst the developing world knows, which Americans consign to our apocalyptic novels—will we truly grasp that *we are animals*. When the world is suddenly or slowly unable to provide, we'll realize: it is all we have, all we have ever had.

Water puts me in this animal place. Beneath the esker-top I call home, the water table lies hundreds of feet deep. Private wells are expensive so we haul water from a community spigot and supplement with rain. Thirty-gallon barrels sit under every eave, gutters providing rainwater for dogs and plants and washing. In a dry spell, I watch water levels in the barrel drop with each watering can or shower. Even though the well is close, down a gravel road, I imagine: if these barrels were all we had, I would be frightened. My own little wetland drying up. The moisture in my skin, my mouth, palpable and temporary. When I have nightmares about climate change, they are dry-mouthed ones.

It is an easy move from my imaginary thirst to a dry world—no, start smaller than that—a dry neighborhood. Eight-Mile Lake, shrunken and shallow. No kettle ponds, no Panguingue Creek coursing fat in spring run-off, tearing out culverts and fragile asphalt. My neighborhood without surface water would be devoid of birds. Imagine it. Spring comes and no robins, no fat orange breasts hollering too early from the tops of stunted taiga trees. No ravens, their heckle and swoop. No swans, no geese in strutting Vs. No cranes. That won't happen, scoff denialists and those skeptical of imagining. No ravens? That'd be the day.

About ravens, at least, they are probably right. Ecological change is more typically incremental than catastrophic. But this is why imagination is critical. It is good to think about what has not happened, even what we think could not happen. Imagining keeps me aware of the high stakes. It reminds me to keep scanning for the birds that are here, to dip my toes into the puddled ditch line. Imagining something lost is a good way to jump-start noticing.

While I value their symbolism and their prodding, I do not believe that nightmares are a resting place. Yes, shrunken ponds and waning birds and too-warm weeks in July and January. Yes, polar bears drowning and villages falling in to the sea. I square myself to these wounds, and I name the injustices they prefigure. But it is possible, in these times, to become afraid of the world, and in the face of that fear, to become numb to the world. There is no more point living in a state of fear—or for that matter, elegy—while we are still alive and thriving than there would be in mourning a person we love before she is gone. When I wake from a dream of thirst, I get a drink and I throw myself into the day, where there is plenty to worry about, but far more to *do*. There is work on behalf of birds—counting, banding, leaving one cranberry on every bush. Work to use less water so that more remains on the surface. Work to replace the words *less* and *scarce* with the word *enough*. Work to notice. To watch.

A critical role is arising: citizen naturalist, "the unpaid, unprofessionally trained non-scientist" committed to noticing.[11] Into the gaps of theoretical modeling, people on the ground step up, counting and watching. Audubon's annual Christmas Bird Count (the 119th, in 2019) is a renowned example of a group of amateur observers aggregating useful conservation data. Thoreau was a citizen scientist long before the label was coined and his observations about the natural world, recorded in his journals, have helped scientists establish climate benchmarks for a time long past.

My neighbors are citizen scientists, too. When I sent out an email asking local residents for observations about our crane neighbors, they sent instinctive hunches, anecdotes and carefully-logged data. Susan said, "There are fewer cranes now, or else more dispersed. We both think they are staying longer." Will noted "they seem to fly over and even circle around but [I've] never seen them on the ground here." Lori has been tracking cranes in her journals for decades: "there were more cranes in the '90s: heavier, steady traffic flow and longer trailing Vs, culminating in a peak year of 1997, when the sky was full of waves of possibly 10,000 cranes per afternoon for nearly the entire month of September. Since then, there has been a steady decline in overall numbers and smaller Vs. The lowest year was about 3–4 years ago, when the swans actually outnumbered the sandhills. This year seemed a bit better. I estimated about 12,000 total cranes during the whole season. A definite decline, but there are always fluctuations. I am hopeful the steady uptick will continue."

These observations matter to scientists, and they matter among neighbors. When we meet at the well or a potluck or out on the trails, we say: *Remember when there used to be Long-tailed Jaegers everywhere? I heard the Great Horned Owls in the aspen grove last week, in April. I haven't seen a Whimbrel at Eight-Mile Lake in years. Do the wood frogs seem quieter to you this spring?* These animated discussions connect us. We want, in the way of most neighborhoods, to keep tabs on the goings on, a kind of community watch. The word *watch* comes from Old English, *waeccan*, "to be awake," and from the proto-German, *wakjan*, which means "to be strong, lively."

Strong and lively reawakens me to the personality of cranes: *watch* and *work* merge with *dance* and *warble*, birds migrating and neighbors staying put. Amidst the shifting ground of climate and projection, worst case scenario and high hopes, the repeated, circling question cranes trigger in me becomes a place to stand. Being brave enough to ask, *what would enough be like*? Letting the mantra pulse beneath all ordinary days: *what the world offers is sufficient.*

Long after they have passed overhead, the cranes remain with me. I picture their pure effort, launching from a soggy patch of cotton-grass into an easterly wind, a thermal lifting their bodies, each the size of a small child. I hear in their elegant music a bet on the future, on the existence of where they are going next. To describe the motions and the sound, my body takes over and as I write, again my arms spread to show how wide the wings, and my throat opens, to mimic that rowdy and exquisite call. Isn't this where any change will originate, any hope? From our animal bodies. From hunger and thirst and the pleasure of eating and drinking, from gesturing limbs, bent legs, movement. From shouts and laughter, warning and song, the indescribable alchemy of effort and the release that trust in the present can bring. All of us together here, trying to find our way.

Chapter 8

The Black and White

Sara Crosby

More often than not, when a bird flies into a window, it doesn't die from a broken neck. Even though it falls to the sidewalk, with its beak bent or broken off, its neck unnaturally twisted, and its head slack, it most likely died of a concussion. Birds, it turns out, have a very sophisticated system of vertebrae, which protects them from breaking their necks; unfortunately, they also have very large brains, protected only by very thin skulls (with the strength of a fingernail in young birds). So when they hit a window with a *thunk*, their brains are shocked by the impact—maybe there's a fracture, maybe there's bleeding inside the brain—and they sail to the ground, landing with a feathery *poof*.

Nationwide, more than one billion birds die after flying into windows each year. That's more than three hundred birds dying per minute during the fall and spring migration months. It varies depending on the size of the dead birds, but volunteers who pick them up estimate that three hundred dead birds is roughly enough to fill up the kind of tall white garbage bag you line your kitchen trashcan with, so that's more than 333,000 kitchen trash bags of birds per year. Millions of migratory birds, who are on their way to Canada's boreal forest for a summer of relaxation and procreation, negotiate with New York City's crop of buildings and skyscrapers each year. The numbers are fuzzy, but easily thousands of them slam into the windows of midtown office buildings because they mistook the glassy reflection of the sky or of Central Park for the real thing. Thousands of dead birds end up lying on the sidewalk in front of Brooks Brothers, waiting, as men and women in business suits swerve to avoid them, for a maintenance worker to sweep them into an upright dustpan with a pert stroke of the wrist.

The dead bird that was in my freezer until recently didn't technically die from colliding with a window. Nor did he die from a run-in with any of the other more popular man-made bird killers, like communication towers, electric transmission lines, or even wind turbines. He also escaped the appetites of house cats, who kill an estimated 100 million birds per year. His problem was a barely visible strip of wire mesh fencing on Charles Street that was rigged to keep rats and other street animals out of an apartment building's yard. He saw some bushes or maybe even a birdfeeder beyond the fencing, took a supreme blow to the head, and was reduced to a little flaccid blip on the concrete.

My father is a bryologist, a collector of mosses, and my mother was the daughter of a bryologist, so the impulse to accumulate things for any kind of study or measurement was not discouraged during my childhood. Our lives were peppered with the run-off of my father's professional collecting. Stacks of homemade paper collecting packets, lumpy and bulging with moss specimens, sometimes sat on the kitchen counter or in the cool of the basement, smelling mulchy, almost edible.

The specimens my brother and I pined for, though, were usually from the fauna realm. We knew if you flipped over a flat stone in the creek by our house, you might get to see a crawfish, reorganizing its appendages in alarm. If you helped to turn the compost pile behind the garage, you'd definitely have to pause periodically to build up your tolerance to, and eventual appreciation for, garter snakes. Beetles were prized—usually chased, kill-jarred, and pinned into a wooden collection box. Voles, dying from feline-inflicted puncture wounds, were inspected, turned over with a stick, and then maybe buried.

In the seventies, my father's office at the Missouri Botanical Garden was relocated to a new, two-story modern building made of dark, reflective glass. Sometimes, while my father sat at his desk inspecting a specimen with his gold-rimmed glasses pushed down to the end of his nose, he'd hear a sudden *whap* against his office window. When he looked up, there was nothing to see, no smudge or smear, an unmistakable sign of an avifaunal window collision. It turned out that the Botanical Garden, despite its mission to preserve, display, and study plants and nature, had a research building made of trick mirrors for birds. My father, seeing the opportunity for a new collecting project, rummaged around in the yew bushes surrounding the building, bagged the bird, and put it in the freezer of his office fridge. He kept bagging and freezing birds for years, and when we as a family discovered, say, a crumpled-up chickadee that had ricocheted off the kitchen window, we always picked it up for my father's collection.

The dead bird on Charles Street was a Black-and-white Warbler, whose streaky body was the size of a jalapeño pepper. I was out walking my golden retriever, Hank, and when I picked the warbler up with a clear plastic poop bag, it weighed as much as a handful of leaves. Other than its rumpled undercoat, which was puffed up in chaotic tufts around its wings, it seemed outwardly unharmed: no jagged broken wings, no chipped beak. Looking at it more closely back at my apartment, I felt very large and awkward in my cargo shorts and bulky sneakers compared to the warbler's needle-thin feet. Its black and white plumage reminded me of the intricate arrangement of notes on a sheet of music. Then, driven mostly by raw childhood associations, I sealed the warbler up in a Ziploc baggy and settled it in my freezer between a pint of vanilla ice cream and a ball of piecrust dough.

On the way to the subway that morning, I called my brother, who was already at his office in Washington, DC.

"Matthew, guess what I found this morning?"

"I don't know, what?"

"A Black-and-white Warbler on the sidewalk."

"Cool," said Matthew, stretching the word out into several syllables. "Did you pick it up?"

"Yeah."

"Awesome. You put it in your freezer?"

"Yeah," I said, feeling giddy, as if I'd captured recess-time and put it in my freezer.

"Dammit, that's cool."

I waited a few seconds, holding my breath through a patch of Waverly Place that smelled like human urine.

"Dude," said Matthew.

"Yeah?"

"You should totally get it stuffed."

Just before my first visit to a taxidermy studio, I was pretty worried about throwing up. My imagination kept riffing on the moment of the first bloody incision and the gooey sound of someone reaching their hand into a stew of organs. What if my stomach couldn't handle it? What if I sat down on a stool beside the workbench, ready to ask questions during the afternoon-long taxidermy project, and suddenly had to scamper off to vomit in the scrap fur bin?

Back then, taxidermy was a mystery to me. Now, after several months of taxidermy excursions, giant projects, like a spread of tapirs scavenging in the jungle at the Museum of Natural History, are still puzzling, but the basic process for birds is clear. A mounting project begins when a taxidermist opens his double wide deep freezer and rummages through the tightly rolled and taped plastic bags until he finds the bird he wants to defrost. He makes a dainty incision in the defrosted bird's abdomen, and, using massaging motions with his fingers, he pulls the skin away from the entire body, pausing to clip the bones that attach the bird's legs and wings. He removes the eyes and the skull, but usually keeps the beak. By then, his hands are lightly gloved in bodily fluids; his fingernails are brownish with blood. He constructs a replica of the bird's body out of wood wool or foam, drapes the skin over it, and adds structure by pegging the bird's new body to its legs, wings, back, and head with wires. He sews the bird's abdomen shut. He works the bird into the pose he wants and adds glass eyes. Finally, with wedding-day zeal and obsession, he arranges the feathers. There's smoothing and petting, tucking away the grey down feathers with the lead end of a pencil so they don't interfere with the more decorative outer feathers, and ordering and reordering each segment of the fanned tail—all tireless, repeated attempts to get the coat of feathers to behave naturally, to fall into place with the fluid effortlessness that really only a living bird can achieve.

That first day, I didn't get sick. I sat on a stool next to Frank Mucha, a mid-sixties Pennsylvania taxidermist who smelled like mint and whose specialty was mounting deer heads. Once he made the first delicate cut into the lower belly of a defrosted Ruffed Grouse, I knew I'd be fine. No liver-colored worms squirreled out at us, and there wasn't a jumble of bloody organs, just the grayish beige of past-prime chicken meat. Frank ran his fingers between the grouse's flesh and skin, gently working them apart down to the rump of tail feathers. As he reached for a stubby pair of scissors to cut the left leg just below what would be the knee on a human, I leaned forward. The crinkle of the disposable plastic apron he'd given me to wear made Frank look up, as if he'd forgotten I was there. "I gotta get a new pair of snips," he said, tilting the scissors in the overhead lamp to inspect their apparent dullness. Then, as Quiet Riot's "Cum on Feel the Noize" came on the nearby transistor radio, Frank nuzzled the scissors into a crevice of flesh, skin, and sticky feathers and severed the grouse's right leg.

Even though the idea of having my warbler mounted and displayed on my mantle was irresistible, I knew it was also illegal. In 1918, Congress passed the Federal Migratory Bird Treaty Act, or the MBTA, which saved hundreds of species from the destruction, and sometimes extinction, brought on by the effects of commercial bird trade. The MBTA essentially bans any human contact with migratory birds. You may not kill, capture, or pursue them; you may not collect their eggs or nests or even their feathers; and you may not pick them up if they are dead. Taxidermists know all this, and so—with the exception of the permitted, but regulated game birds from hunting—they will not get near a migratory bird.

I was interested in keeping things legal with my warbler, at least while I weighed the option of doing something illegal, like stuffing it. I called the Fish and Wildlife Service in Albany, where a woman named Joanne talked me through the possibility of getting a salvage permit, which lets people temporarily possess a migratory bird. However, the salvage permit application wanted legitimacy I didn't have: it required me to include authorization from the scientific or educational institution that would receive any salvaged specimens I collected when I was finished studying them.

As a workaround, I joined a team of local Audubon Society volunteers and walked a weekly route, scavenging for window collision victims. The program was called Project Safe Flight, and my orientation packet included a piece of paper stating that I was covered under the Audubon Society's salvage permit. The packet also included all the items I would need to scour the sidewalks on my Wednesday morning loop around the World Financial Center: a map of the route that indicated all its high-risk glassy areas, rubber gloves, Ziploc bags (gallon-sized for woodcocks, sandwich-sized for warblers), a Sharpie, a dark green hankie for picking up injured birds, and several hole-punched paper lunch bags for transporting injured birds.

In the early morning each Wednesday, the World Financial Center broiled with people. Mercantile traders, sporting prep school-style jackets and the pallor of stress, huddled in the forty-degree weather to smoke before the opening bell. Ferries from New Jersey disgorged men whose belts were lined with communication devices and women with umbrellas and soft coolers filled with lunch. I was most concerned with maintenance workers and security guards, though. Since they were the only people who shared any interest in the concrete that made up my route, their presence made me anxious. The first couple of Wednesdays, I spent the half-hour walk to the Financial Center hoping that I would not find any birds; then, on the walk home, after having found nothing, I'd feel guilty—what if there *had been* birds but the maintenance man had swept it up just before he power-sprayed the Winter Garden plaza? Then, I started making friends with both maintenance men and security people. I handed out Project Safe Flight brochures, which I kept along with all my other paraphernalia in various pockets of my raincoat. I shook hands, gave them my cell phone number, and talked about dead birds. Once, a red-faced security guard named Chuck showed me where he'd seen several dead birds earlier in the week. As we walked along the side of 2 World Financial Center, he pointed at the knee-high hedge beside us.

"Whenever I see them I usually just sweep them into the bushes," he told me in a thick New York accent. "People complain otherwise. They see the bird and it's horrible to see, and then they look at me like they want me to bring it back to life." When I asked him what the birds looked like, or how big they were, he shook his head and smiled. "These birds, you can tell, aren't from around here. They're too good-looking to be from around here!"

At the end of the spring migration, we dozen or so Project Safe Flight volunteers found only seventy-two dead birds after eight weeks of walking four different routes throughout Manhattan. The project's manager said the total was unexpectedly low and was probably due to a cool spring and delayed foliage: if the trees that surround glassy windows aren't leafy, birds aren't interested in them, real or reflected. Despite the low numbers, the project's data confirmed what the Audubon Society has long known—some of New York's most familiar buildings also kill the most birds. The glass room that houses the Egyptian Temple of Dendur at the Metropolitan Museum of Art overlooks Central Park and therefore kills lots of birds. But Project Safe Flight's biggest killer was a U.S. Postal Service mail processing facility on Ninth Avenue in Chelsea; during the fall 2006 migration, one volunteer remembers collecting thirty-three dead birds, mainly woodcocks, from the building's sidewalks. (After years of negotiating, New York City Audubon convinced the U.S.P.S. to install screens over the facility's windows, which ironically are purely decorative and have nothing behind them but concrete. The building's number of bird kills is now almost zero.)

I was lucky. I didn't find any dead birds. I did do a weekly double-take at the same disintegrating orange peel near the north entrance of 4 World Financial Center. And I did find a Song Sparrow who had slammed into the same window twice; I used my green hankie to bag him and then taxied him to a veterinary volunteer at Animal General on the Upper West Side. Her name was Rita, and she was wearing her horseback-riding outfit. The sparrow spent two days recovering, and then Rita left me a voicemail saying she'd released him in Central Park and that he hoped to never be back.

The project was a great distraction, and being covered under the Audubon Society's salvage permit helped me feel a little less crooked about the little plastic-bagged warbler in my freezer. Sometimes I'd squint into its baggy on my way to the ice trays, again relieved that it wasn't swept up and sent to the indignities and disgrace of a landfill-bound garbage bag. But the door of our New York apartment's freezer wasn't all that respectable either. I wanted the warbler out in the open, displayed like family photographs or hydrangeas from the garden. I wanted guests to be able to study it, as they studied the bound spines on our bookshelves.

In 1803, when one could still see thousands of White Pelicans on the Ohio River or forests heavy with millions of Passenger Pigeons, John James Audubon arrived from France. He was eighteen and quickly became known for his fidgety energy and rough English. He'd already started drawing birds in France, but in America he began to think of his work as more than just scientific profiles of lifeless specimens. He became focused on trying to bring birds to life through his drawings; he wanted to permanently maintain the beauty of birds in their natural setting in an artful way, or, as he put it, in a way that was "pleasing to every person."[1]

The problem with this, of course, was that in order to get a bird to pose for his sketchbook, Audubon had to go out and shoot it, which created his lifelong challenge of trying to portray the beauty of a living bird while looking at a very dead bird. In his *Ornithological Biography*, the scientific essays that accompanied his *Birds of America* collection of prints, Audubon recalled that as a nature-obsessed child he'd begun to recognize that "the moment a bird was dead, however beautiful it had been when in life, the pleasure arising from the possession of it became blunted." Throughout his life he grew more skilled at combating this problem: he developed ways to quickly pose a freshly killed bird using a wooden board and wire; he always tried to make his subject life-sized, no matter if it were a warbler or a pelican; and he often used poses that made it seem as if the birds were making eye contact with the viewer.

Dave Tuttle, a bird taxidermist in Susquehanna County, PA, is somewhat of a modern-day version of Audubon. He has Audubon's same restlessness, graying wavy hair, and slightly receding hairline. He still uses names left over from Audubon's time for certain birds, as many hunters do: my House Sparrow is his English Sparrow; my female Long-tailed Duck is his Oldsquaw Hen. And, like Audubon, Dave Tuttle loves to shoot guns. He likes to sit in the woods in a full-body camouflage suit that's designed to make him look like a tree branch. He likes the sport of shooting, the years of tradition behind hunting, and the yap of his dogs working. Especially, though, he likes birds.

"I really love birds, man," he told me one day in his studio, "because they're moving fifty miles an hour." He leaned closer to a vice-gripped mount of a wood duck, inspecting what looked like plaster casts on its feet. "So many people deer hunt, but with deer hunting, it's just so boring. You just walk out and start shooting. With birds, I like that you have to consider their habitat and their food and their mating habits."

I was sitting in an office desk chair across the wooden workbench from Dave, watching and listening and trying not to lean back too far or else I'd get my ponytail tangled in the steel brush of the bird flesher on the counter behind me. "So," I said, as Dave switched from the duck's feet to adding a strip of blue painter's tape to its fanned out tail feathers. "You're out hunting for a lot of the same reasons I go out with my binoculars."

Dave laughed, which, like all his laughter after what sounds like years of smoking, turned into a phlegmy cough. "Yeah," he said when he'd recovered. "Kind of."

It's obvious from Dave's studio's collection of owls, waterfowl, and wild turkeys in permanent perching and soaring poses, that he also shares Audubon's desire to bring dead birds back to life. Once, as he watched a pair of Eastern Phoebes, Audubon reported looking "so intently on their innocent attitudes that a thought struck my mind like a flash of light, that nothing after all could ever answer my enthusiastic desires to represent nature, than to attempt to copy her in her own way, alive and moving!"[2] Audubon made thousands of sketches and outlines of birds, relentlessly practicing how to express on paper not just what they looked like, but how they moved and how they behaved when they were alive. Whether you love birds or not, Audubon's work is awesome: by so artfully documenting actual birds in their true surroundings, he does manage to keep them, metaphorically at least, alive. (This is especially true of the now extinct specimens he portrayed, like the Carolina Paroquet or the Passenger Pigeon.)

Dave Tuttle's work is also exceptional. The first time I visited Dave's studio was in March 2007, and it was filled with Wild Turkeys. I stood in front of one of them, a male in his strutting position, and thought of the time my husband and I were hiking in Westchester County and met the calm leaf-crackle of a flock of Wild Turkeys. We were on our way to see some migratory waterfowl, so I just kept walking, thinking that we could enjoy the turkeys as they cautiously moved out of our path. We'd taken about twenty steps and were watching the flock do their prehistoric slink down an incline to our left when we heard a loud gush of air and saw a gobbler fully strutting about fifteen feet from us, looking directly at Mark. Tail feathers fanned, wings pointing toward the ground, red wattle jiggling, he was perfectly posed, like the stereotypical turkey that's tirelessly depicted among pilgrims and Indians in Thanksgiving motifs. If it had suddenly raised its foot to turn showily to the left and then the right, the strutting turkey in Dave's studio could have been the same bird.

Dave's birds were so seductively real that I often caught myself thinking he was magical, like he had some god-like power to recreate birds. But his talent has a very simple, hard-earned source. He's hidden in the woods for years, watching birds. He knows their motions, their posture, their feather coloring and arrangement, and their mannerisms so well that he can effortlessly reconstruct them in his workshop. He's also compulsive about detail. As his assistant Bob stood next to the workbench tossing half a dozen skinned-out House Sparrows in a pail of corn-cob grit to dry them, I watched Dave sculpt the sparrow bodies, starting with an arm's-length block of polyurethane foam. It took him at least forty-five minutes of fast-paced work to produce one sparrow body the size of his thumb: he took measurements; he sawed; he took more measurements; he shaped the foam with sandpaper to get its exact size and shape, periodically stopping to run his fingers over the body's developing shoulders; he made and attached the bird's neck, a cotton-covered wire bent to create a perching pose. When he was finished and passed the body to Bob so he could begin fitting the skin, Dave didn't take a break to smoke or chat before starting on the next one.

But Dave's obsessive quest to recapture nature's beauty in a mount was an unnatural and brutish process, and the way taxidermy toggles back and forth between dogged natural beauty and brutal lifelessness began to increasingly disorient me. If Dave were mounting, say, a Ruffed Grouse—his favorite bird to hunt—it would be unrecognizable for half the taxidermy process. I've never seen one alive, but a male Ruffed Grouse has a large black ruff, or feathery collar, around his neck; his posture is erect and proud, and his movements pertly exhibit his ruff and his turkey-like tail plumage. After skinning it, Dave turns the remaining coat of skin and feathers completely inside out, and it becomes an off-white lumpy mess similar to a mound of the spongy flatbread you're served in Ethiopian restaurants. When the skin is flipped so the feathers face outward again, it looks like roadkill—matted and oily and flat. And it garners the same response as roadkill: whenever I saw skinned out birds, particularly larger birds, I had an automatic urge, probably brought on by the revulsion and shame of mutilation, to just look away.

When they are finished, Dave's mounts seem more like pieces of art, objects to be admired—exactly what I had imagined for my Black-and-white Warbler. I, like Dave and like many hunters, had wanted to have a perfectly preserved bird perched on my mantle to resurrect its natural beauty. But now I was too familiar with taxidermy; regardless of Dave's ability to ultimately restore a bird's natural beauty, I looked at a mount and saw only a very—almost exhaustively—dead animal. I couldn't imagine my warbler ending up like the dozens and dozens of birds I'd seen in taxidermy workshops: those mounts convey the belief that there is an abundance of birds out there, that they are such a commodity that they should become merely decorative, that their glass-eyed stares represent nothing other than their status as beautiful hunting souvenirs.

When my father's accumulation of dead birds started to crowd his reserves of coffee beans, he called a colleague at the Smithsonian's National Museum of Natural History. Then he packed the dead birds in a Styrofoam chest of dry ice and shipped it to the Department of Vertebrate Zoology. Someone from the Division of Birds sifted through the dead birds and chose to add several to the museum's collection. In the spring of 1974, according to the Smithsonian's online database, my father is credited with collecting an Ovenbird, a Ruby-throated Hummingbird, and lastly, a Black-and-white Warbler.

Around the time I gave up on the idea of taxidermy, my Black-and-white Warbler was cocooned in an ice crystal crust in its Ziploc bag. I called my uncle, who directs the Florida Audubon Society and who, after jokingly telling me to have it stuffed, said that his organization didn't have any use for a Black-and-white Warbler from New York City. Then I found Kim Bostwick, curator of birds and mammals at Cornell University's

Museum of Vertebrates. When I arrived with my warbler in a soft cooler of ice over my shoulder, Kim, with the deliberate calm that reminded me of my father on his moss collecting trips, reached for a notepad on her desk and wrote down the bird's history, mainly exactly where and when I'd found him. There was a tiny refrigerator under one of her lab tables, and she knelt down to seat my warbler next to several containers of soymilk. It happened fast, while I was re-zipping my cooler. The refrigerator door shut with a muffled thump, and I wondered if that would be the last time I'd see the warbler. Then Kim, who has light brown curls that, after a couple of days in the field, develop into a seventies-glam frizz, took her fistful of keys and asked, "So, do you want to see the collection?"

The museum's collection consists of about fifty thousand study skins—dead birds, which have been skinned out and stuffed, but are not realistically posed. They lie, belly-up in their wafting preservative scent, in the paper-lined drawers of giant metal cabinets that are sealed up against hungry moths and beetles. During my tour, Kim rolled out drawer after drawer of specimens to illustrate all the different uses for dead birds. A drawer of dozens of grosbeaks was an easy lesson in hybridization, since some of the Black-headed Grosbeaks had streaks of a Rose-breasted Grosbeak's red-rose plumage. ("That's what we're trying to do," Kim told me. "We're trying to catch evolution in action.") Two trays of tanagers from Mexico, Peru, and Columbia had feathers that were shocked with color: swatches of aqua blue and indigo, a yellow the color of acorn squash, the kind of bursting, toxic-seeming green I'd always assumed could only be made by chemicals. These were not just to humble me and my drab human coloring; they showed off how study skins defy age. These feathers were brighter and more pure than any color I'd ever seen, and yet the identification tags tied to their feet reckoned they had been preserved in 1908 or earlier. Another drawer, laid with four puffins, brought up the collection's ability to help the research world: scientists in Maine, for example, can study the levels of chemicals puffins absorb, by harvesting comparison data from the specimens' feathers.

We visited a cabinet filled with pelicans, and Kim pulled out a Brown Pelican so I could closely inspect the accordion folds of its beak's pouch. We leaned over a Wandering Albatross that was collected in 1849 that couldn't be picked up because his head was a little wobbly. A bottom drawer at the end of one of the aisles of cabinets held a Philippine Monkey-eating Eagle, who, lying stiffly on his back, was as big as a seal. We stopped by the cabinet of extinct specimens and saw a Passenger Pigeon, a Carolina Paroquet, and a pair of Ivory-billed Woodpeckers. The Passenger Pigeon, which stood erect, as if it were listening for danger, had clearly come to the museum as a part of a taxidermy mount. When I noted this as we walked back to her office, Kim said, "It's funny. I just never know how people are going to react when they look at all these dead birds."

She told me that visitors often thought of the collection as a bird morgue, which was understandable: no matter how well they held their color and overall shape, these birds had been prepared to look lifeless, with wings and feet folded away and nothing but white cotton showing through their eye sockets.

"As a person who's in charge of a lot of dead things," she told me, "I don't want to put people off. For some people the specimens make it more real, and it's such a valuable experience. And some people look at a dead specimen and they just see a dead thing."

For Kim the specimens are both lovely and symbolic. Back in her office—under the relentless gazes of an antique Kakapo mount and Kim's enormous flat-screen Macintosh—we sat in sliding desk chairs. Kim explained that when she sees a dead bird in the collection and especially during fieldwork she can't appreciate its beauty enough. "It's just so magical, so moving," she told me. She looked out into the woodsy area beyond her office window. Then she pulled up the cuff of her coarse khakis and scratched her ankle. "That's why our collection seems more like an altar or a shrine where we show reverence to nature."

Before I left, Kim took my warbler out of the refrigerator and unwrapped him. We stood at one of her lab tables, and she tilted him in my direction between her hands, holding his beak between her thumb and index finger. His eyelids were open, and his black eyes seemed to be squinting. I *wanted* to leave him with Kim, who I knew would so thoroughly appreciate him, but my body felt a little bit saggy, as if I needed a nap.

"I'll make sure one of our best students gets this guy," Kim said. It might take a couple of months, she explained, but ultimately my warbler would be scientifically prepared and then join the thirty-five other Black-and-white Warblers in the drawer marked with their scientific name, *Mniotilta varia*. "Don't worry," she told me, wrapping him back up in his cling-wrap cushion. "When the student sees this bird, he'll recognize how beautiful it is."

Chapter 9

One Single Hummingbird

Eli J. Knapp

Science and everyday life cannot and should not be separated.
Rosalind Franklin

Why, I wondered, had I opted for two low-clearance minivans for cheese-grater roads like this? Sure, the keyless entry and automatic doors were nice. As was the discounted price. But the manufacturers at Chrysler obviously hadn't been considering accessibility to Lucifer Hummingbird haunts when designing these. My current predicament could not be pinned on shortsighted auto manufacturers, however; *I* was the one who had led us here.

For three weeks we'd studied birds in Western New York. Then, to better appreciate bird diversity across the U.S., I'd brought my twelve college students here to Texas for the culminating week of my intensive ornithology course. It was a trip to gain perspective, make some memories, and chase a few southwestern specialties. The most special of all, I preached in the days leading up to the trip, was a dainty hummingbird sporting a slightly decurved bill, a forked tail, and a shimmery purple throat, a feature I ceaselessly reminded my students was better called a gorget in ornithology circles. We'd spent the previous six days watching Greater

Roadrunners, Phainopeplas, and Elf Owls. The little Lucifer, several of which could be comfortably tucked into a shirt pocket, was our final quest.

The dusty, potholed road—more resembling a cattle trail—bumped my thoughts around. Bird love flowed in my veins; I wanted to behold each and every species. Now as a professor, I wanted my students to do the same. But why was I so focused on this one particular species? Sure, Lucifer Hummingbirds were pretty. But with the exception of the Chihuahuan Raven, so were all the birds we'd seen so far.

In field guides Lucifer Hummingbirds didn't seem too different from the dozens of Ruby-throated Hummingbirds we'd seen in the East. Nor were its proclivities different. Like the other twenty-four hummingbirds that spend time in the U.S., the Lucifer specialized in drinking nectar. More noteworthy was its name. Lucifer means "light-bearing" in Latin, likely a reference to its iridescent gorget. That, and according to one dubious source I really want to believe, a group of Lucifers is called an inferno.

Driving me, down deep, wasn't behavior, appearance, or funky nomenclature; it was rarity. Only 10 percent—the intrepid few—entered the U.S. each summer, sporadically appearing in New Mexico, west Texas, and little slivers of Arizona. Most stayed in the stronghold, Mexico. Yes, rarity alone, that fickle and somewhat arbitrary characteristic shared by snow leopards and tanzanite, had pulled my carload, and the second minivan of students I could now no longer see in the dust cloud, to this specific spot.

The spot, named Christmas Mountains Oasis, seemed an odd place for a bird with devilish nomenclature to frequent. Nothing outside the windows resembled the yuletide season. The boulder-strewn landscape had remarkably little vegetation, just a mosaic of brown and gray baking in the noontime sun. An intermittently updated blog had led me to the property's owner, Carolyn Ohl-Johnson. Permission and directions followed a pleasant phone call. "Sure you can come!" Carolyn had said cheerily. "I'll be gone, but stay as long as you want!" Now, however, it felt like a fool's errand. Clutching the wheel ever tighter, I was feeling more and more like the Grinch. I couldn't steal this Christmas Oasis, but I could certainly call off the quest.

"The other car is gone!" Kyle yelled up from the back.

"Gone?" I asked incredulously.

"Yeah, it's nowhere back there!"

I hit the brakes, our dust cloud enveloping us like it does Pig-Pen from *Peanuts*. How could I have lost the other car? It had been behind me just a minute ago. We hadn't been going faster than ten miles per hour.

"Should we go back?" I asked, hoping for a balanced democratic decision. Silence. Finally, Kyle, the group's self-elected spokesman, nodded. One ugly five-point turn later, we were on the narrow road headed back. Just as abruptly, we found—and nearly collided with—the other car. Andrew, slightly older than the other students, had generously agreed to be my second driver. I'd had him in several classes over the years, and we had a good relationship. Now, as he pulled up alongside, I worried that I'd strained it. He put his window down. His shaky voice and shiny eyes confirmed what I suspected: he wanted nothing more to do with this ill-fated trip.

"I don't think this van is fit for this," he said, tapping his steering wheel with both hands.

"It can't be much farther," I offered optimistically, wanting to cash in on my poor investment. In truth I had no idea how much farther it was. But I was unsure how far I could push this group without inciting a mutiny. "Can you go a little more? Or should we try the rest on foot?"

"On foot," Andrew replied without hesitation. Several heads nodded agreement from the seats behind him. We pulled the minivans over and piled out, one girl inadvertently spraying sunscreen in her eyes as we readied ourselves for a forced march in the blazing sun.

A half hour later, we found much needed shade under faded awnings lovingly placed alongside a series of recently filled hummingbird feeders. We had found it! Christmas Mountains Oasis was exactly that, an oasis. With water features, feeders, and skillfully cultivated vegetation, Ohl-Johnson had turned a parched patch of Texas desert into a verdant stage for wildlife; an avian Shangri-la. In addition to a lava colored coachwhip that slithered past our feet and an oversized leopard frog guarding a pond, the several-acre spread hosted vireos, buntings, kestrels, quail, and, not more than a few minutes into our hummingbird vigil, a pair of shimmery Lucifer Hummingbirds, the male cut out for Mardi Gras.

Hummingbirds delight in a way other birds can't: their disproportionate bill size, cool disdain for gravity, and unbridled curiosity in the particulars of their environment. Our brightly colored clothing led them to inspect us like a fastidious grandmother looking over a grandson. Some students ducked and dodged as the hummingbirds zipped past, instinctually avoiding the whizzing projectiles. Logic assured us otherwise; the birds' superior navigational system would never allow mortal collisions with bipeds.

We ogled the birds, walked the grounds, and couldn't resist snapping a class photo in front of the feeders. The lack of a grand backdrop was irrelevant; what mattered were the starlets who surrounded us, and the sacrifices we'd made to get here. Our satisfaction was visceral; we had made it somewhere few others did. An ebullient walk and a bumpy bit later, we were back on the highway, joyful survivors of our encounter with Lucifer in a part of the U.S. that could certainly double as Hades.

The course ended a few days later, and we were back at our college in New York, where we'd started the class and prepped for Texas. My students dispersed to summer jobs and myriad responsibilities. Me? I had the mind-numbing chore of collating receipts and grading assignments, one of which was for every student to keep a daily field journal. Unmotivated to organize receipts, I turned to the stack of journals. Top of the pile was Kyle's. I paged through, skimming his entries and assorted data he'd compiled. Then I paused.

"Continuing on," Kyle wrote on one of the last pages, "we drove down a crazy dirt road to find one single hummingbird and that was quite the experience." I reread Kyle's run-on sentence. Something about the way he phrased "one single hummingbird" stopped me. Technically, of course, Kyle was wrong. We'd seen at least a half dozen hummingbirds on that trip to Christmas Mountains Oasis. But I understood his sentiment; we'd risked flat tires and fragile psyches all to see one single species. A species weighing less than a nickel, and one that before the course started was unknown to my students. Although he didn't write it, the question Kyle asked was obvious: *Why?* Why risk so much for something so little?

The more I've hung out with seasoned birders, the more I've realized they rarely ask themselves this question. Most have reconciled themselves to their passion, and besides, they already know the answer. Birds may be small and weigh little. But they're big in everything else. Abilities. Beauty. And, most importantly, the way they consume our imagination like few creatures can. As a biology professor in midlife, I'm content with my interests. Like I said, I was born this way. But I have struggled to articulate why I like birds—and why I like to share my passion—to others. A passage in *Care of the Soul*, a book by Thomas Moore, has helped me of late. "If you don't love things in particular, you cannot love the world, because the world doesn't exist except in individual things."[1]

During the portion of the course that occurred in Western New York, I'd taken my students to the rim of the magnificent 500-foot-high gorge at Letchworth State Park. The park, often called the "Grand Canyon of the East," is worthy of the epithet. To better awaken the senses, I love to teach on location, using nature's unpredictable props as my PowerPoint. In mid-lecture, a winged missile shot past us. With that kind of blazing speed, it could only be one thing: a Peregrine Falcon. Most peregrines do just that, streak by like a shooting star and dissolve equally quickly. Not this one. It flew in a U shape, gracefully arcing up against our side of the canyon wall. It alighted on a small ledge twenty yards to our right. Our bird's-eye view could not have been better.

Despite regular, sometimes weekly visits to the park spanning well over a decade, I had never seen a peregrine in the gorge. Nor had I seen one in my county. The peregrine clung to the cliff as we watched, its periodic cry piercing the cool canyon air.

I watched the peregrine intently. But I couldn't resist watching my students, too. Some were content with a snapshot and a few seconds through binoculars. Others were transfixed, field glasses affixed to faces seemingly with glue. These few I knew were locked in the falcon's grasp, spellbound. For this long moment, they were peregrine watchers. Despite our rapt attention, none of us watched peregrines the way a British recluse named J. A. Baker once did. In his small, unadorned book, *The Peregrine*, Baker wrote a simple sentence that stuck out to me as Kyle's had. "For ten years," he wrote without embellishment, "I followed the peregrine."[2]

Ten years! During the ten-year span Baker followed the peregrine, he recorded 619 peregrine kills. *Six hundred nineteen times* he watched a peregrine dispatch a lesser bird and then documented the remains. Me? I've yet to find my first. Nor have many other naturalists I know. But Baker's book extends well beyond a keen eye and a fawning appreciation of nature. He wasn't a mere naturalist. As Robert Macfarlane acknowledged in the book's introduction, Baker, of whom history knows precious little, aimed not just to watch peregrines. He was after something far bolder: he wanted to become one. "Wherever he goes this winter," Baker wrote,

"I will follow him. I will share the fear, and the exaltation, and the boredom, of the hunting life."[3]

In the pages that follow, he does exactly that, tracking the peregrine pair for miles and miles, day after day. This is where *The Peregrine* becomes as singular as the falcon Baker follows, where it separates itself from all other natural history books. Macfarlane suggests that Baker succeeded in becoming a falcon. Or came as close as any bipedal human ever has. He points to Baker's pronoun usage as evidence. Subtly—but ever so importantly—it shifts. When Baker discusses the falcon midway through his book, "I" changes to "we."[4]

Baker's pursuit reveals the face of devotion. I've long envied this trait, largely because I lack it. My interests span too widely. As much as I'd like to follow falcons, I'm too distractible, too much a generalist. Sure, I like birds. Really like them in fact. But Baker's devotion requires forsaking all else, something I'm unable, and unwilling, to do. While single-minded devotion is out of my grasp, appreciation isn't. This I have in abundance. So much so that passing it on to my students is as natural as breathing.

Letchworth's peregrine provided the perfect opportunity. Nature had given me a three-pound prop of duck-killing fury. The fastest bird in the world. Now was not the time to conclude my ornithological lecture; the falcon would. It was time for silence; to let the bird—and the moment—speak.

Speak it did. Binoculars remained up. Phones remained stashed. Chatter ceased. The falcon continued calling. Nervous Rock Doves clung to cliff faces, their liquid eyes scanning the vertical walls for the source of the cry that upset their world. Far below, Common Mergansers blasted away upriver, leaving a wake of white water behind them.

Neither I nor my students have the ability and attention span to follow individual falcons for years on end. Nine-to-five jobs, grocery runs, dental appointments, grass to mow—mundane daily practicalities dictate our lives. For all of us not named J.A. Baker, moments are all we can spare.

The good news is that moments—with their particular places and particular species—do suffice. Henry David Thoreau alluded to this with his famous line "In wildness is the preservation of the world."[5] By "wildness" (often misquoted as "wilderness"), Thoreau was referring to the fields and forests around his cabin, the ordinary wildlife, species he regularly encountered on his walks and out his windows. Some scholars have suggested that Thoreau's ability to celebrate wildness around his little Walden Pond allowed him to appreciate wilderness on a larger scale.

I agree. My concern for the imperiled ecosystems of the world would be zilch if I wasn't daily moved by the wildness in my own backyard. Newborn fawns trying out their legs for the first time, skunks palpitating my pulse as I concede them the back deck, even the scraggly young robins begging for worms on crisp May mornings. These are everyday species. Everyday moments. But without them, life would be blander, colorless even. "Teaching a child to care for a goldfish," wrote author Sallie McFague, "learning about its needs, respecting its otherness, delighting in its shimmery colors and swimming skills—is a better education in caring than is a lecture on global warming."[6]

With all the ecological problems facing us—habitat loss, land fragmentation, pesticide use, invasive species—the time to champion the particular seems ripe. For most of us, attachments to particular species lead us to larger, more global concerns. This is why, for the days I have left to teach, I will continue carting my students to places where peregrine interruptions are possible. It's why, poorly chosen minivans or not, I'll keep leading them down dusty, potholed roads. I'll use the props nature provides and seize each moment, no matter how trivial. I could settle by giving each student a goldfish. But I'd rather gamble; push them outside and bring them somewhere well off the beaten path. Maybe one day they'll mutiny and demand to stay in places with Wi-Fi and phone coverage. But if they don't, the reward may be as memorable as it is dazzling: a rarely seen species sporting a decurved bill, a forked tail, and a shimmery purple gorget. Hopefully an inferno. But even one will suffice. One single hummingbird.

Chapter 10

Assault on the French Canal Bridge

R. A. Behrstock

It all began in 1881. Count Ferdinand de Lesseps, flamboyant French diplomat and financial promoter, had just begun supervising work on what promised to be the next engineering wonder of the world—a sea-level Panama Canal. The financial rewards would be staggering; an inland channel spanning the Isthmus of Panama (then Colombia's province of Panama) would connect the shipping lanes of the Atlantic and Caribbean with the Pacific Ocean and eliminate the long and occasionally disastrous 8,000-mile voyage around South America's Cape Horn. Such a grand scheme might have been dismissed as the ravings of a mad man; but de Lesseps had already earned international fame by completing another monumental engineering project—the Suez Canal. Linking the Red Sea with the Mediterranean, this inland waterway allowed shipping companies transporting goods between Europe and the Far East to forgo the long trip around South Africa's Cape of Good Hope. In March 1878, Colombia granted France a concession to begin construction of the canal, allowing them eighteen years to accomplish the project. Engineers estimated seven to eight years to complete the work. Count de Lesseps, now elderly, was sincere but overly optimistic, predicting the canal's completion in only six years.

After two years of aggressive fundraising, de Lesseps's dream of another canal was finally becoming a reality. Investments totaling hundreds of millions of francs rolled in from both sides of the Atlantic, funding men and machinery for their assault on the Isthmus. Work began in earnest in 1882. But after six brutal years, France lost its battle to dig a massive trench through the steaming tropical rainforest. Landslides propelled by 200 inches of rain each year continuously refilled the huge channel. The labor force, comprised mainly of Creoles from the West Indies, was decimated as upwards of 20,000 workers died from yellow fever, malaria, typhoid, snake bite, and unspeakable accidents. Many attributed the failure to engineering issues, among them, the stubborn insistence of de Lesseps (who was not an engineer) on digging at sea level, coupled with his disdain for powerful locks that could hold the ocean at bay. By 1889 work on France's canal was terminated. In 1904, the heavy equipment was sold to the United States, which succeeded in constructing a canal by traversing the hills of the Isthmus via elevated waterways and three sets of huge locks. In the wake of de Lesseps's failure, political scandals revealed widespread financial improprieties, including the bribing of over 150 French deputies, many of whom were tried in court. The French middle class, heavily involved in funding the project, suffered a crushing financial blow. Count de Lesseps himself narrowly escaped a prison term, paying a huge fine instead.

It would have been instructive—perhaps even entertaining—to consider those events 100 years earlier that led to this evening's circumstances. But the thousands of shots crackling from the dark forest around us made it difficult to focus on the historical perspective.

In Vietnam, they called it "Illum"—soldier's jargon for an illumination flare. It was a tube of white-hot burning magnesium powder that descended from the sky, buoyed by a miniature parachute. Armed with a mortar, a soldier could fire such a flare 700 feet into the air, or hundreds might be dropped from a slow-flying airplane, fondly referred to as "Smokey the Bear." For the thirty seconds it drifted downward, its flame turned nighttime into day, and anyone caught in the open knew they were a target.

And there we were—not in Vietnam, but birding in Panama—bathed in the eerie white glow of Illum, and pretty much scared shitless.

February 20, 1981. I was coleading my first birding tour in Panama. For the prior twelve years or so, I'd been pursuing Neotropical birds through Mexico and Central America during school breaks and summer vacations. Recently, I'd left my fisheries job at Humboldt State University in Northern California and moved to Houston, Texas, where Ben Feltner, Linda Roach (now Linda Feltner), Mary Ann Chapman, and I formed Peregrine Tours, with Ben at the helm. Ben had recently dissolved his former company, Merlin Birding Tours, and we were keeping up the momentum with fresh partners and new itineraries.

Now, a week or so into Peregrine's inaugural Panama trip, we were birding on the Caribbean Slope and encountering many birds we'd not seen earlier on the Pacific side. It had already been a productive day that began with a morning hike on Achiote Road—then one of the country's premier birding sites. Achiote was a bird-watcher's dream: little or no traffic, tall roadside trees full of raptors, woodpeckers, parrots, toucans, trogons, flycatchers, and my favorite, the huge White-headed Wrens whose chortling calls rang from the highest branches. Stepping into Achiote's shaded understory invariably produced a surprise—a flock of Black-chested Jays mobbing a boa constrictor, a lemon yellow eyelash viper, or a pair of shy Gray-cheeked Nunlets roosting in a vine tangle. After lunch at the American Legion's restaurant (where our waitress was astonished to learn that we'd all come to Panama *just* to look at birds) we continued to Ft. Sherman. The fort's woodlands overlooked the western bank of the Panama Canal, and were an important center for our government's jungle operations training program.

By all accounts, the day had been a great success; everyone had seen many "lifers," including some of the region's iconic birds, such as Purple-crowned Fairy, Broad-billed Motmot, Black-breasted Puffbird, and Long-tailed Tyrant, plus a dozen antbirds whose identifications were keeping everyone on their toes. Remaining at Ft. Sherman through dusk, we extended the afternoon's birding with a night drive back to the Washington Hyatt in Colón. We'd timed our return to put us on the little-used road in darkness, hoping to encounter a Great Potoo, the largest member of its nightjar-like family; a beautiful Black-and-white Owl, whose barking calls might alert us to its presence; or perhaps a nocturnal mammal such as an olingo or kinkajou searching for figs or nectar-rich flowers high over our heads. Apart from seeing a two-toed sloth and hearing the occasional mournful wailings of Common Potoos, it had been an uneventful evening.

Up before dawn, we'd had a very long day, as often happens on birding tours—especially those with enthusiastic new leaders. Some of the participants fought to stay awake as we drove southward, straining our eyes and ears in the mostly unbroken forest bordering the mangroves of Manzanillo Bay. David Markley, a young naturalist aspiring to work with us and later hired as a leader for Questers Tours, was in front of me, driving one of our two Japanese rental cars. Next to him was Polly Rothstein from Purchase, New York. Polly, a fiery political activist and a life-long advocate for women's reproductive rights, became a good friend and a veteran of many tours to the tropics. Behind them sat Marg Benson, a member of the National Parole Board of Canada. I was following, driving the second car. Ellis and Ann Van Slyke from Grosse Pointe, Michigan, were in the back seat. Ben, the tour's primary leader, was sitting on the hood of the front car, holding a Q-Beam—a powerful spotlight that plugged into the car's cigarette lighter—standard birding gear back then. As we inched along in the darkness, Ben swept its beam slowly through the canopy, then up and down the trunks of prominent trees, watching for the eye shine of a bird or mammal while dodging the huge beetles and moths that were attracted to the light.

By then, all that remained of de Lesseps's French Canal was a narrow, water-filled slit through the forest, no more than two miles long, perhaps 200 feet wide, terminating at its south end in a small lake. Its location and pitiful ending invite cruel comparison, as it is but a stone's throw from the mighty Gatun Locks finished by the Americans in 1913 at the Caribbean entrance to their new and wildly profitable Panama Canal.

As we rolled onto the bridge that spanned the French Canal's dark waters, Feltner's eyes caught a reflected glint high on his left. Gesturing at Markley to stop the car, he pointed the Q-Beam into a tree bordering the canal. In that moment, we were all treated to a preview of the wrath of God. Innumerable shots rang out from every direction, fiery tracer rounds flew past us, and the entire bridge was bathed in a harsh white glow as burning magnesium flares descended around us. Shots were fired so fast, it sounded as if a frustrated army was emptying its weapons as quickly as possible. The countless bangs and pops suggested that guns of various calibers were being fired. More comfortable with birdcalls, I wondered if these noises were to be my final auditory determination as a tour leader—or as a human being? As I desperately tried to disappear under the steering column, I glanced toward my back seat. Somehow, Ann and Ellis had managed to wedge themselves under the floor mats, and I still remember being envious of their questionably defensive position. Terrified, we all thought we were done for. Ben, bravely, remained perched on the hood of the front car, waving his arms and yelling, "Cease fire, cease fire," into the surrounding darkness. Perhaps he'd realized we were being attacked with blanks. Or maybe he just hoped they were lousy shots.

Within seconds, both cars were surrounded by heavily armed U.S. soldiers, their faces streaked with camo grease and their bodies burdened by full jungle combat gear. Simultaneously, my participants and I realized we were still breathing—as much surprise as relief.

A young officer, clearly distressed that one of his trigger-happy privates had ruined his platoon's maneuver by engaging an unknown target, strode up to Feltner and barked:

"What the f*ck are you doing here?"

Feltner calmly replied: "We're looking for owls. What the f*ck does it look like we're doing?"

About that time, I noticed that the soldier closest to my window had his rifle carelessly pointed toward my head.

"Troop," I shouted more loudly than I should have (my bravado enhanced by the monstrous adrenaline rush we were all experiencing) "get your *&%$# weapon out of my face!"

Sheepishly, he adjusted his sling and stepped back. About then, a visible sense of uneasiness swept through the young warriors as they collectively arrived at a couple of unfortunate conclusions. First, they had defended the bridge with an all-out assault against two Nissans full of birders. Second, one of the privates had fired upon civilians and the rest of the patrol had followed suit in a grand and wasteful manner. Certainly, there would be hell to pay after we drove away.

On December 31, 1999, forty-five years of jungle warfare training ended when the Panama Canal and the U.S. military bases in the Canal Zone, including popular birding sites such as Ft. Sherman, Ft. Clayton, and Albrook Air Force Base, were transferred to the Panamanian government. Until then, birding tours in the Canal Zone often encountered U.S. forces on jungle maneuvers. Even when we didn't actually see soldiers, rope, wire, spent flares, and other bits of hardware littering the understory revealed their recent presence. Occasionally, the evidence was more dramatic, as when a rocket arched over the forest, setting off a chorus of roars from the scattered bands of mantled howler monkeys. Or, just as we'd tracked down a Jet Antbird or were standing in the middle of an army ant swarm, studying a woodcreeper's field marks, a convoy of several dozen military vehicles would roll by, scattering the birdlife in all directions. Walking through the woods along Achiote or Black Tank Road, we would encounter soldiers crouched in the undergrowth. Most were young privates from Wisconsin, Tennessee, Nevada, or some other faraway place, and some hadn't been in Panama much longer than the participants on my tours. Sometimes they'd just wave us by—hoping we wouldn't reveal their position. Occasionally they'd ask if we had seen "the aggressor," assuming that we would have discerned "the aggressor" from any other U.S. soldier. Communicating with outsiders, birders in this case, seemed to be part of their strategy, and I remember asking one private if it was OK for me to provide information on the troops they were hunting or evading. I should have expected his response when he winked at me and said: "All's fair in love and war."

After a number of tours in Panama, I took a job with another company, so it was about twenty-five years before I returned to Panama as a tour leader. Things had changed a great deal. Panama City was bigger and wealthier with new high-rises crowding the shore of Panama Bay. There were more roads, more tourist lodges, and a beautiful new bridge crossing the Panama Canal. For better or for worse, the Canal Zone and thousands of U.S. military buildings had been turned over to the Panamanians. The birding was still wonderful, but I missed the chance encounters with U.S. troops, their soft voices whispering at me from under a clump of heliconia leaves asking: "Have you seen the aggressor?"

Chapter 11

Wild Swans

Alison Townsend

The Tundra Swans arrived the day before Thanksgiving, descending like a band of luminous angels to tiny Island Lake, its cattail marshes rattling with the first chill of winter in Wisconsin. Navigating by the stars and their memory of earth's moonlit landscape, they came, traveling from their summer breeding grounds in shallow pools, lakes, and rivers in the Arctic toward their winter residence in Chesapeake Bay and the marshes of Virginia and North Carolina. White as alabaster, warmed by swansdown the color of snow that has not yet fallen, the great birds swept in, accompanied by their pearl-and-silver-plumaged young, who were making the journey for the first time. Resonant with the mystery of the world's great migrations, they arrived, and because they seemed to appear out of nowhere, it was hard not to see them as the messengers that angels are said to be—saints, perhaps, or the spirits of ancestors—descending through layers of black toward their own faint reflections, the lake's surface glinting, dark as tarnished mirror. It was the time of year when the veils between worlds are said to be thinnest, when one walks about feeling spiritually porous. What could we learn from this unexpected visitation?

Not that we saw them arrive, mind you. A quarter mile away, nestled under the quilts in the north-facing bedroom of our house on the drumlin hill, our daytime selves suspended in a starry net of dreams, we did not witness the splash and din of their arrival. We did not (at least at first) hear their rich bugling cry—*whoo-hoo, whoo-hoo, whoo-hoo*—that is neither honk nor coo, but something wilder and stranger. We did not see them set their wings, stretched as wide as six feet, for landing, gliding in over the black water. We did not even know they were present, settled in as if they had always been there, until the next morning when, out in the fields with the dogs, my husband saw them, drifting like dreams on the dark blue water at the farthest side of the lake. "What are those big white birds?" he told me he wondered, mistaking them at first for the Snow Geese that also visit these waters, before it occurred to him that they might be swans.

A trudge back up the hill for binoculars confirmed his intuition. They were as real as the shape of my own fingers, adjusting the binoculars to see them more clearly after my husband pulled me, half-awake, down the hill, under the barbed-wire fence, and across the fields stippled with fallen cornstalks, broken cobs glowing here and there like dull brass in an otherwise sere landscape. "They're on the lake," he cried. "Hurry! There are swans!"

And there they were, the distinctive black beak and the small yellow spot at the base of the upper mandible (like a decorative bead of yellow jasper or a tear made from sunlight) identifying them as native Tundra Swans, not the Mute Swans I'd glimpsed now and then at a park or zoo while growing up in the Northeast. Necks held straight up, the swans drifted and preened and fed on wild celery and arrowhead tubers, perhaps recalling the manna grass and marine eelgrass of summer. Tundra Swans, which used to be called whistling swans for the sounds their wings make in flight, often travel in groups of several hundred. According to my worn *Audubon* guide, "they present a spectacular sight" when they make mass landings in places like the Niagara River. Unfortunately "they are [also] sometimes swept over the falls there to their death."[1] Our group of swans was small, though, about two dozen, and, given the number of smoky gray young, seemed to be composed of several family groups traveling together. Like adolescents not quite ready to leave home, cygnets remain with the parent flock for at least a year, learning the route and where to feed and rest.

Drifting like a flotilla of small boats—did the sight of swans inspire the first sails?—the great white birds seemed a kind of miracle we'd stumbled on by accident. It had been a difficult season. My sister's husband had been diagnosed with cancer. My writing mentor and dearest friend was experiencing severe pain and the ravages of age. I was feeling more worn down than usual by my work teaching young writers, and not particularly thankful for a life that is, in many ways, blessed and abundant. Life is always harder than we think it should be. But it is ours, isn't it? And here were these magnificent birds, sailing along on our lake, going about their business and filling me with an awe that knocked me sideways and took me outside my small human concerns. Bound by cycles of seasonal change and patterns of birth and renewal, the sight of the swans comforted me on some essential level, offering what I can only describe as the solace of wild things. If they could manage to do something this enormous, guided by star patterns and earth's magnetic fields, I could navigate my life, couldn't I?

Gazing at the swans, I thought about something Rachel Carson said in her acceptance speech for the John Burroughs Medal: "If we have ever regarded our interest in natural history as an escape from the realities of our modern world, let us now reverse that attitude. For the mysteries of living things . . . are among the great realities."[2] Migration is surely one of them. Looking at the swans put my own life in perspective, allowing me to see it as the speck it is, compared to this greater, shimmering whole I was privileged to witness.

As we watched the flock, a cob and pen set forth, moving from one floating sedge island to another, their three cygnets strung between them like a garland. They paddled and drifted, paddled and drifted again, seeming to move effortlessly through the still landscape, their reflections following them through the water. I thought about what it must be like to live this way, stopping partway through a journey of over 3,600 miles to rest for a day or two until the wind changes and something—instinct, perhaps, that thread of innate, blue *knowing*, or simply the urge of a winged creature to be in the air again—pulls them aloft, sometimes to heights of 9,000 feet, and onward. I imagined them, spiraling back up into the darkness in a whirlwind of feathers, like some heavenly host, a feather or two dropping behind them like white leaves. I could see how, as my husband said he'd read, they might spook a driver whose path they happened to cross.

Studying the swans as they drifted on the lake, like music made visible instead of occurring as sound, I understood why they have engaged my imagination. Swans often appear in stories having to do with transformation, as if something about their beauty and elegance suggests a hidden life we all contain. I remembered how, as a child who never felt like she fit in anywhere, I identified strongly with "The Ugly Duckling," a story that seemed to offer the hope that, though shy and awkward, I, too, might possess some secret interior beauty, not unlike the Inner Light my Quaker-educated mother spoke about. At night, I read from my mother's many books of fairy tales, entranced by images of swan maidens and swanskins and the seven brothers who were turned into swans, saved only by their industrious sister weaving them flax shirts that made them boys again. When I was seven or eight, I fell in love with Tchaikovsky's *Swan Lake,* which I discovered while working my way through my parents' leatherette-bound edition of the *Standard Treasury of the World's Classical Music*, which my mother purchased, one record a time, at the A&P. I'd drop the diamond needle on my father's RCA Victrola, listening past the initial hiss and scratch for the music that summoned me to be a swan maiden, dancing on an icy lake. When I'd fallen ill at my grandmother's big house in Philadelphia, I loved how she tucked me up in her own bed, the swan lamp on the table beside the bed left on all night, as if it were flying beside me, keeping watch.

Oddly, given the place they occupied in my private mythology, I can hardly remember ever seeing a swan in the wild. Once, just months before my mother died, we kids were taken to a lake somewhere in southern Westchester County, where we ate ice cream and watched Mute Swans sailing across the sunlit water, their necks curved gracefully, proud as the figureheads of swans boatbuilders carved for good luck. My father took us there that winter to go ice-skating, and I skimmed over the frozen surface, numb inside, wondering if the swans we'd watched were trapped somewhere under the ice. Later, when he remarried too quickly, I remembered reading "The Children of Lir" in my mother's book of Irish fairy tales. In that story, a wicked stepmother turns her children into swans and sends them away for 900 years. Would something like that happen to me, I wondered, or would I just have to keep washing the kitchen floor ten times until I got it right?

It was only as an adult, when I traveled in England and Ireland, that I saw swans as it seemed to me they were meant to be seen. I spent an hour one September afternoon sitting beside the lake near Ann Boleyn's childhood home, Hever Castle, near Edenbridge in Kent, watching swans skim across the water. They seemed infinitely sad and stately when juxtaposed to her unhappy story. In Ireland, I saw a pair of swans on a river in Kildare, shortly after praying at Bridget's Well for healing. They seemed to mean something numinous, though I didn't know then how prominently swans figure in Celtic legends. According to James MacKillop in *A Dictionary of Celtic Mythology*, they are often depicted as "the epitome of purity, beauty, and potential good luck," as well as symbols of "communication between the Otherworld and the world of mortals."[3]

But as my husband and I watched the swans on our lake, I was absorbed only by the immediacy of their presence, not what they might represent to me imaginatively. All I could do was stand, with my back against one of the oaks that fringe the lake, drinking in the sight, thinking what my father once told me—swans mate for life. If one bird dies, he'd said, the survivor often will not mate again. I knew my father was thinking of my mother when he spoke. Following her too-early death, he remarried not just once, but twice, each time with disastrous results. Late in his own life, he admitted that in each case he'd been trying to replace her, telling me she was the love of his life. If only he had behaved more like the swans he admired and apparently really was, I thought. It might have saved everyone in our family a lot of pain.

My husband took photographs, including one of the largest swans (a cob?) standing high on its black-stockinged feet, neck stretched high, chest out like a schoolyard bully, flapping its enormous wings. Looking on, I thought of how fierce swans are said to be, attacking intruders and defending their mossy, mounded nests. Later that afternoon, as our collies loped alongside the Yahara River a few miles away, we looked up and saw a small flock of a dozen swans, flying in the direction of Island Lake. Were they "our" swans, or part of another group? We could not know, of course, though we liked to think they were ours.

Recalling Yeats's swans that "suddenly mount / and scatter wheeling in great broken rings / upon their clamorous wings," I listened for the whistle of their wings over the water, but could not hear it.[4] There was nothing but the sound of the river running beside me and the sight of the swans themselves, gilded by the light, merging with the clouds they so resembled. The day after Thanksgiving the wind shifted and the swans were gone, vanished as mysteriously as they had come. Without them scattered across its surface, like the cotton ball clouds fallen from a child's painting or the white lotuses that bloom there in summer, the lake looked empty, forsaken, waiting.

The next time we walk outside at night I will ask my husband, who knows the stars better than I do, to point out Cygnus, the stellar swan, its wings outstretched above us, sailing down the Milky Way, the River of Heaven. Some say Cygnus represents Orpheus, who was changed into a swan at his death and placed near his magic harp (Lyra) in the sky. But as we look up, our faces faintly illumined, as he traces the celestial bird's outline, I'll think of the Tundra Swans. I'll think of how they sojourned so briefly beside us, delighting us with their purity, elegance, and mystery. A part of me will want to know what the swans' appearance "meant," pressing it for some sort of message. But in the end, seeing them will have been enough, and I will be thankful for that. One should not, after all, question the holy too closely, though it is good and right and proper to recognize it as what it is. Still, I'll wonder where the swans are now, hoping they are in a marsh or coastal lowland, and that we will see them again on their way back to the Arctic in the spring, this small body of water in Wisconsin encoded in their memory now, the way they are in ours. I'll think of them all winter, they whom the Greeks made a symbol of their Muses, they whose name comes from the meaning of the word "song."

Chapter 12

The Snowy Winter

David Gessner

The best moment came when I was standing alone out on Coast Guard Beach—on almost the exact spot where the naturalist-writer Henry Beston lived for a year in his outermost cabin on Cape Cod's ocean shore—when the young Snowy Owl rose off the tundra-like marsh with a black duck in its talons. The duck hung down limp below the owl, and below that the duck's lifeless feet hung down even lower, like damaged landing gear. The owl flew over the dunes with its prey, wanting to be alone, and I felt something of the irritation it must have felt toward me when I saw the couple climb over from the beach and attempt to follow it. We both, owl and man, were no doubt experiencing some variant of the same basic thought: What are they doing on *my* beach, on my *territory*?

My irritation faded when I caught up to the couple. Bundled up like refugees against the cold, the man and woman were considerate, giving the owl plenty of space. They were not birders, didn't even have binoculars, but they were so delighted by what they had seen that I had a hard time being grumpy. The woman's face was radiant. "Radiant" is a word

I had been using a lot during my last few days of owl-watching, though mostly to describe the white unworldly shine of the Snowy Owl's feathers.

"It's only the second time I've seen an owl in the wild," the woman said.

She acted as if she had witnessed a visitation, which she certainly had.

"They're amazing birds," the woman said.

I told her I couldn't agree more.

We kept a lookout for another flight up from the dunes but decided to give the owl a wide berth as it continued its repast of uncooked duck.

It is late January 2013, and I am now fourteen days into my twenty-three-day stay on Cape Cod. I am here to finish a novel that I began to write over thirty years ago when I was living a couple miles to the east of where I am now. "Taking a book off the brain" was how Melville described the last draft of *Moby Dick*.[1] This was a process that took Melville a period of months not decades, but I'm hoping for something similar, in technique if not result. In a sense I am back here to finish a job. A wild job, but a job nonetheless.

Since moving down to North Carolina over ten years ago, we have not gotten back to Cape Cod nearly as much as I'd like. But we have lucked into a great house-sitting gig in the town of Brewster near Slough Pond. We stay in the home of Katy Sidwell, an artist, extravagant and generous, and Steve Sidwell, a former defensive coordinator for the Patriots, Seahawks, and Saints. While the Sidwells are away, we walk Buddy, the rescue dog they found in New Orleans after Katrina; watch their art-filled house; feed their cat(s); mow their lawn (in summer); soak in their hot tub.

I have been here two weeks, and have stuck to a schedule that I have found pleasant but that I understand others might find insane. Buddy and I get up around three, and after I feed him, I start typing away in the basement, Buddy usually lying under the desk. It has been a long time since I have spent full days writing to the exclusion of all else. I was worried I might go crazy being here alone, and while I still might, I am excited about the big writing days that yawn ahead.

What has lifted this trip into the unexpected are two gifts from above. The first is the owls. As it turns out, I am not the only one who is spending these winter months on Cape Cod: an irruption of Snowy Owls has occurred, which means that one of the most beautiful, stunning, and startling of birds, formerly rare in the Lower 48, has descended on us in numbers that are at once thrilling and puzzling. At West Dennis Beach, one of the most popular Cape owl-watching spots, you don't even have to look for the owl when you get to the parking lot; you just look for the line of cars of people there to see the owl. There are times when the crush is too much, when the owl lands in a tree near the parking lot, for instance, and hordes of photographers try to muscle each other out, not content when their photogenic owl—remarkably tolerant when not chewing on a duck or other prey—allows them to get as close as fifteen feet, but flies away when they insist on five.

But most of the people I have encountered have been as delighted, and as respectful, as the woman back on Coast Guard. And while the birds are fairly easy to find, watching them for any amount of time is something that must be earned. That's because the owls are not the only thing that has traveled down from the Arctic.

Snow has been the second gift. More snow than I remember from all my time on Cape Cod. It started the other morning when I was bird-watching in the hot tub, staring up at three Tufted Titmice when a Cooper's Hawk came diving after them. As I was looking up something came floating down, a flake landing on my cheek as I steeped in the boiling water. That was the first flake of millions. Before leaving North Carolina, I had thrown my cross-country skis in the back of the car at the last minute. Now, each day after I finish writing, Buddy and I push off up the driveway, and down the back road to the woods through the abandoned summer camp. The snow is great, with a good base, and keeps getting better.

It is also perfect weather for owl-watching. The other day in West Dennis, for instance, as the flakes fell at the start of our second major storm, I watched with a couple other people as an owl alighted on an Osprey nest that had been built on top of one of the blue boxes for trapping greenheads that you see on the marshes here. Standing atop the huge nest, only six feet off the ground, the owl appeared to be wearing a shining white robe, though the robe was flecked dark with brownish markings the color of chips on a cinnamon scone, likely marking it as a female. As she swiveled, her great disc of a face transformed itself, one moment a white lion, the dashes of black for bill and eyes a mere cartoon outline, the next more hawklike and predatory as she lifted a handlike talon to scratch herself, and then, when her eyes turned entirely away, a featureless white dome. Mostly she squinted in the wind, but when her eyes re-opened they shone yellow. Amid the sere colors of the marsh her whiteness jumped out, though the next day, with the blizzard's help, her camouflage would work perfectly.

Her tail feathers blowing behind like streamers in the heavy wind, the snowy seemed made for spectacle. She had *presence*. While not our tallest owl, Snowy Owls are our heaviest, their thick feathers necessary for insulation. Unlike most owls, snowys are diurnal, which means that in the Arctic summers they have twenty-four hours for hunting, and that means that patient humans can watch them hunt during the winter. Among other fascinating facts about the owls, the Cornell Lab of Ornithology reports that images of Snowy Owls have been found in cave paintings in Europe. This didn't surprise me: after several days out on the tundra-marsh, my own journal pages have filled with sketches.

For the last three days I have not been able to get my car out of the driveway and have not been desperate enough to use Steve's truck. I have plenty of deli turkey meat for sandwiches and cereal and milk and other basic provisions. Buddy has been a great companion and he sleeps curled into my legs. When I was skiing in the woods this morning I looked into his brown eyes and couldn't believe how smart and aware (I will not say "how human") they looked. I was not all that surprised during our drive the other day when he reached over and lowered the electric window on the passenger seat side, then stuck his head out. I put the window up again, and he just looked at me, reached over, pushed the button, and lowered it again. This went on a few more times until I finally had to put the child lock on.

There is a new twist to my daily routine. At the sea camp there is one cabin with a back deck that cantilevers out over Herring Pond. It has just one room, likely a playroom in summer, and is mostly boarded up, but if you peek around the boards you can see a foosball table and a bed frame without a mattress. That is where I stop and take my skis off halfway through my ski, where I take the nearly frozen beer out of my backpack, and then pace back and forth in the snow of the deck, continuing my morning writing by dictating into my microcassette player. I am sick of typing by that point in the day, and as Churchill was always quick to say, "A change is as good as a rest." Snow covers the ice that covers the pond, and lately I have seen coyote prints weaving through the snow. Whenever I get close to the edge of the pond I hear the Great Blue Heron as it goes *sproaking* off. Once I finish the last cold sips of beer I toss the empty under the cabin, where I plan on collecting them all when the snow melts.

I am excited about the book I'm writing but I have a slightly fearful *Flowers for Algernon* vibe going too: I know that I need to finish it before these three weeks are up or I never will. So much of my adult life is about interruption, so these three weeks are both precious and, of necessity, obsessive. Melville said of that deep writing state that it was the "strange wild work." That's where I want to go.

If I have my own white whale, it is this novel itself. This is the same book, more or less, that I began right after college, the first year that I spent a winter down here. I lived in my family summer home, a house that wasn't winterized. I started the novel in November and gradually went crazier throughout the winter. It is easy to romanticize solitude and isolation, but all too quickly they can lead down a very dark road.

The first time I saw a snowy, and the only time I saw one before this year, I did not go in search of it. Back then, when I still lived on Cape Cod, I went for daily off-season walks on the beach, and that was all I was doing on that morning more than fifteen years ago. When I got to the end of the path to the beach, there it was. It stared at me with black slits for eyes, the ocean wind blowing its white feathers back like a boa. It was less than ten yards away and didn't feel like moving, so it just sat here, looking both spectacular and perfectly at home on a beach that on that frigid day must have truly resembled the tundra where it spent its summers. I stared for a while, then walked away. For the owl it was no big deal, but for me it was one of the great days of my birding life.

Yesterday morning I came upon a crime scene during my ski up to the wintry summer camp. Two regular features of those walks have been coyote prints on the ice and the Great Blue Heron. Today the two things came together in bloody fashion. A headless heron lay just off the path with clear evidence of the culprit: bloody footprints leading away and a circle of urine to mark the area of the kill.

This morning to confirm the victim's identity I returned to the overlook, and walked out onto the ice to a spot where someone had been ice fishing. No sign of the neighborhood heron.

This morning I walked back to the spot, and the heron was gone, head and all. There had been another nighttime visitor to the site, however: raccoon tracks this time as the secondary predators moved in.

I took some pictures and walked home, and soon Buddy and I were in our positions, me at the desk and he below. The novel is going well, though I worry I am running out of time.

I didn't know anything about writing a book when I started the first incarnation of this one so long ago. I just knew that it took time and space—*uninterrupted* time and space—and I had a sense that once that year was over, and what my father loved to call "the real world" started crowding in, those things would be taken away. I didn't know for sure what the next years would bring but I was pretty sure it wouldn't be vast stretches of free time for brooding and creating.

The golden retrievers signaled the end. On Cape Cod most of the beach houses sit abandoned all winter long; when I walked the beach I always had it to myself. Then one day two friendly golden retrievers trotted up, and I knew their owner wasn't far behind. I also knew in my gut that winter was over and so was the first of my wild isolated times.

It's true that I didn't immediately get married, have kids, and take a job as an insurance salesman, and that compared to many people, my life remained fairly wild. But that spring, and the warm weather, did mark the end of something. I was emerging out of the raw wilderness of beginning.

For me the story of the snowy's invasion is a story of wonder and wildness. But this story of course takes place in our modern, troubled world and so doesn't quite have the innocence and simplicity of fairy tale. The owls have made a great splash, in both social and regular media, a Beatlemania for birds. But the Beatlemania metaphor works only to a point. When the Fab Four landed at JFK to launch their first U.S. tour—almost fifty years ago to this day by the way—they were rushed by frenzied teenage girls. When the Snowy Owls landed at the same airport, they were greeted by men with guns, hired by the Port Authority, who shot and killed three of them, worried that they would interfere with air traffic. (Airports, like beaches and marshes, have that low, flat treeless look that to the snowy mind says tundra.)

Then there is the bigger question, the ten-thousand-dollar question, which is, *Why* is this happening this year? There are plenty of theories for the influx of birds, though no one really knows for sure at this point. During the first days of my own owl quest, I made a decision to *not* read up on the subject. This was partly an attempt to have more of an unpremeditated experience, to perhaps re-create the spontaneity and wildness of that first unexpected owl encounter I had long ago on my beach. I wanted to see owl, experience owl, maybe even if I was lucky *be* owl for a second or two, before I read owl. But of course I was sometimes with other people when I watched, and so the question of *why* naturally came up. From my fellow watchers, I learned the word "irruption" and heard that there had been a boom in the lemming population in the Arctic, which in turn led to a boom in the owl population, and that the young birds were coming south in search of new territory. But of course this didn't answer the real question of why we were seeing something that was unprecedented over the last hundred years.

Maybe another reason I didn't bone up on the bird at first is that I have read, and written, enough environmental stories in recent years to know where they all end up heading. How nice to have a hopeful story, a story of humans rushing out to take joy in encountering other animals, a story that doesn't end with the doom of the planet, brought about, as always, by the evil done by the hand of man. The modern environmental mind naturally, and understandably, runs toward self-castigation. Isn't it nice to have a story that doesn't?

So I felt a little sag in my chest when I started doing research and heard that our old friend climate change was being dragged out as a possible culprit. It makes sense of course: Couldn't warmer temps in the Arctic lead to the explosion of rodents, and that lead to the explosion of owls?

As the popular eBird site puts it, "These owls are surely telling us something, but we still don't understand exactly what."[2] It will not be terribly surprising if the answer to that mystery is an ugly one. John Schwartz, writing in the *New York Times*, quotes Cornell ornithologist Kevin J. McGowan on the disturbance of the snowy's Arctic environment: "That has to be one of the most vulnerable ecosystems on the planet. That's going to be one of the first places that falls apart when there is warming in the atmosphere."[3]

It may turn out that the arrival of these radiant visitors is yet another sign of the end of the world. Maybe so. But for this morning at least I choose to think otherwise. This morning I will again bundle up and get out into the cold with them, keeping my distance while watching, tuning out the tolling of the usual bells of doom, living for just a moment in a world other than human and seeing their arrival here as a generous visitation, an unexpected joy, an undeserved but well-appreciated gift.

CHAPTER 13
KOAN
URSULA MURRAY HUSTED

It was a long time before I tried to fly again.

Nights were spent alone by Lake Wingra listening for Short-eared Owls and whip-poor-wills...
...and days, after and before work, drawing on the University of Wisconsin terrace, watching ducks raise their ducklings and pelicans stopping to rest on their journeys north.
When asleep, I dreamed of flying.
One afternoon, the radio said that the big pelican migration would be coming through Horicon Marsh. I called in sick, something I never did, and drove northeast on U.S. Highway 151.
It was my first true solo expedition.
I waited up all night in the Visitor Center parking lot with my sketchbook ready, too excited to sleep.

When the sun rose it glinted off the brash ice on the surface of the marsh. Red-winged Blackbirds chortled and coots and mergansers dove and darted around the polynya.

That's a nice drawing.

Thanks! I'm waiting for the pelicans.

Sorry, but they came through yesterday.

Oh.

Your wings are the ones given to you and they can be hard to use.
That's not something that goes away when you get older.
It is better to crash than to fear flying.
I mourned the loss of my pelicans, but I had taken a risk and drawn what I found.
Walking to my car, I looked up and a flock of Sandhill Cranes dropped like pieces of white paper from the sky.

Chapter 14

This Is My Tribe

Thomas Bancroft

I yank my coat collar up higher around my neck and pull my wool cap down over my ears. It's twenty degrees, and the light breeze makes it cold even with thick layers of clothing, but the landscape has me stopping regularly to stare. White ice crystals cover the vegetation in millions of delicate jewels, and they glisten, sparkle, and flash in the early morning light. The sun, just barely above the eastern horizon, gives a warm cast to the vegetation and hoarfrost.

I shuffle along this western Washington trail, and suddenly a flit appears behind some shimmering gem-covered leaves and then flutters out into the open. A tiny Ruby-crowned Kinglet moves faster than a gymnast on high bars, jumping from branch to branch, peering under leaves, hovering briefly in front of a curled white-wrapped leaf before moving on. The movement is almost frantic but not unusual for this miniature bird. Then a second one pops up from behind another weed. Both are only inches off the ground, working feverishly through the sparkling foliage, on a mission. Normally, they forage high in the canopy of trees. The hoarfrost must be making food more available in these weeds. These small olive-green birds with strong white wing bars and white eye-rings don't seem fazed by these low temperatures. I pull my collar up tighter, drawing my head into the soft down wrappings as I hop from foot to foot, watching them dash one way and then the other.

The first time I saw a Ruby-crowned Kinglet was in the spring of my sophomore year in high school. Some birding buddies and I had gone to Presque Isle on Lake Erie to look for spring migrants. Often during the third week of May, thousands of northbound birds congregate on this thin peninsula before making the overwater flight across the lake. My friends wanted to see as many spring warblers as possible. Many of these species only pass through Pennsylvania from their Latin America winter homes to Canada's boreal forests where they breed. In the spring these birds are brightly colored, singing, and looking like they are ready for the senior prom. We had left Pittsburgh two hours before dawn so we could be there for the morning feeding frenzy.

John, our leader, was a lot like me: a skinny kid, without much muscle, and of similar height. But he was a senior, and even though only two years older, he seemed to know everything. John had seen in his life more than 200 species of birds, a feat I couldn't ever imagine matching. He talked about birds and the natural history of western Pennsylvania. His voice sounded like a professor, and he had an air of authority about him. He seemed to know his plants, too, and told us stories of looking for salamanders, including the hellbender, an aquatic salamander I had never heard of. All of his seemingly limitless knowledge gave me hope that I could learn, too. I wondered if he was shy in school, like me, without many friends. His descriptions of warblers, salamanders, and his outings in western Pennsylvania were simple and straightforward, not like Charles.

Charles, a junior, was heavy-set but seemed strong, determined, and energetic. Each time John described something he had seen, Charles had a story about something that was better; he'd seen more Snow Geese and a Blue Goose, too. He had found a rare warbler, the Brewster's, a hybrid of the Blue-winged and Golden-winged. He always had seen more individuals, a closer look, or in some other way one-upped John. Dave, the last member in our car, was in my biology and geometry classes, so I knew him better. He was quiet, too, didn't say much on the drive to Presque Isle, but had clearly been birding for several years with people who had shown him lots of species.

My car mates all seemed to have the field marks in their heads for more bird species than I even knew existed. My mother had liked birds and encouraged my interest in them and had given me a pair of good binoculars the previous Christmas. But here was a group of people more experienced and excited about my passion than I'd ever seen. Most boys I knew in middle school and then high school focused only on sports—football, basketball, baseball—and girls. I never told them that I liked birds. As the chatter buzzed through our car, I couldn't sit still; I'd found my tribe.

At the park, we jumped out of the car and started down a narrow dirt trail into a thicket of deciduous trees. New light green leaves kept twitching in the imperceptible breeze, making me stop and put my binoculars on each movement—just another leaf. My group began to move away, and I hurried to keep up, not wanting to miss anything.

Then, just above eye level and along the left side of the trail was a definite bird. I jumped with excitement but still managed to find it with my binoculars as it flitted from branch to branch, checking under leaves, on the twigs, for things to eat. This little olive-green bird, only a few inches long, must be one of the warblers we had come to find. "Hey, guys, I think I have a warbler here, but don't know what kind." My mentors ran back along the trail, all finding the bird instantly. "Oh, it's a Ruby-crowned Kinglet!" They turned immediately to head back on their quest with hardly a glance at my find. My shoulders slumped, and the air seemed to leave my body, but this bird seemed incredible to me, so delicate and yet active. I continued to watch the kinglet, mesmerized by its energy, how it flicked its wings with each hop, how it could move so gracefully, an acrobat.

Many more kinglets appeared throughout our search for warblers, and I continued to think they might be warblers, alerting my friends again. They started to ask, "Does it have wing bars and a white eye-ring?" If I said yes, they didn't pause in their quest for warblers. Finally, John tapped on my upper arm and said, "Don't worry about asking," as he pulled out the bird book. "We all had to learn the first time. Here, see, these are the key field marks for kinglets"; he pointed to the eye-ring and wing bars. He then turned toward the back of the book and flipped through the warbler plates. "See, these don't have the wing bars and eye-rings at the same time and see how their body shapes are a little different, foraging behavior, too. It takes time, you'll get it, keep asking. We don't mind"; and he closed the book and hurried down the trail, adding, "Let's find some warblers."

I couldn't believe my good fortune in finding John, a person who loved sharing birds with a novice like me. Brightly colored spring warblers were abundant that day on Presque. Black-throated Greens with their yellow cheek patch, green backs, and black throats, Black-throated Blues with their white bellies and blue backs, Chestnut-sideds with their brownish stripe along their sides, Yellow Warblers with their bright yellow bodies and red pinstripes, and so many more. Their songs sounded like soloists playing flutes, oboes, or violins. Each bird was dazzling, thrilling to see and hear, making me hungry for more. Our group considered it an outstanding day. We saw almost two dozen warblers, plus tanagers, grosbeaks, flycatchers, sparrows. For me, though, each sighting of a Ruby-crowned Kinglet seemed remarkable; this drab olive-gray bird, smaller than the warblers, would send a chill through my body, a thrill like Christmas morning when I was small. Something about it seemed special to my naive teenaged eyes.

Now, almost fifty years later, I still marvel at these little birds. Those friends were amazing, too. Although I've lost contact with them, I remember them as if it was yesterday. The kinglet in Washington did an acrobatic roll and grabbed a morsel. Smaller than warblers, but yet they stay here in Washington—and a few in Pennsylvania—for the winter, while all those warblers head to Latin America. Kinglets weigh less than two small green grapes. These in front of me today have their feathers puffed out, making an airtight covering much like a warm down quilt, but still, in this cold, they will need lots of food to fuel their furnace. They need about ten calories a day, the equivalent of five raisins. It takes a lot of arthropods—dormant spiders and insects and their eggs—to feed these little birds throughout the winter. Right now, these kinglets are focused methodically on fueling that grape engine, the most important thing for survival in this cold. I am always amazed that these guys don't move farther south each fall.

Here in Washington, wintering kinglets migrate down from high-elevation spruce-fir forests where they breed or from Canadian breeding sites, to live through the cold months in the Puget Sound lowlands. As I jump from foot to foot to stay warm, they make me ponder my decisions on where to live. Friends back East ask me the same question. How can I stand staying here, in the rain and gloom of the Pacific Northwest? These little birds come only this far so in the spring they don't have as far to fly to reach their breeding grounds. One hovers below a branch for the longest time, its wings going in a blur as it sticks its minute bill between frost-covered leaves. His scarlet crown patch shows briefly, a male. He disappears into the dense weeds.

I came to Seattle to start over after my wife died. I found a good job, a girlfriend, but then the job disappeared and the girlfriend, too. Most friends seemed to vanish at the same time, but a few stayed tight. My relatives, all in Pennsylvania, ask if I will move back East, but I seem to stay here. The kinglets dart among the bushes, madly searching. The hoarfrost crunches under my feet as I mince after them. These little birds are my tribe, too.

Several times, I've promised myself that I would decide what's next; in September after spending the summer hiking, in July when I finished the writing course, or after the next backpacking season. Yet I continue to stay, and those decision points come and go. There is something unique, mesmerizing, about this place, the rain, the vegetation, the animals, the landscape. The resonance is strong; this country gives my body courage. These little kinglets epitomize active, energetic people focused on life. They have a laser focus on their needs.

During the same year of high school as my first sighting of a kinglet, I had a gym teacher, call him Mr. Bob, who badgered me during PE. Even during an afternoon gym class, he wore no sign of a five-o'clock shadow. I pictured him shaving between classes. His biceps bulged in his T-shirt, and his thighs looked hard as steel. His sneakers were so white, he couldn't possibly have ever taken them outside. His face, though, was almost hollow. When I was trying—in vain—to do my push-ups, he would come down on one knee, lean over toward me and say, "Come on, Bancroft, can't you do more than that?" His neck muscles flexed in and out, in and out, in and out. Then he'd stand back up, grunt a few times, and walk away.

Though I never once was able to lift my chin above the bar, Mr. Bob stood right beside me, his shoulders rising and falling, biceps flexing and flexing, as he spoke right into my ear: "Come on, Bancroft. What do you do on your farm anyways?" I never told anyone that I lived on a chicken farm, least of all Mr. Bob. Most of my high school classmates came from an affluent suburb of Pittsburgh, and one of their favorite derogatory comments was to call someone a "farmer." Sick throughout my childhood, allergic to many common foods as well as suffering from chronic asthma, I had never developed the muscles of a normal boy, and by the time I got to high school I felt like a weakling, embarrassed by my lack of strength. Some, mostly the jocks, snickered in the background when Mr. Bob got on my case; others would say to me in private, "Don't let him bother you." I endured Mr. Bob, the jocks in PE, and pursued birding with my tribe.

Here in Washington, the ruby-crown in front of me dashes along a wintry branch, wings twitching with each move. It peers left, then right, before hanging upside down on a side stem to pick a morsel from the underside of a leaf and gulp it. Many of the curled frost-covered leaves are as big as the kinglet. Its white eye-ring makes it look like it's wearing spectacles. Its eyelid repeatedly blinks as it twists to glare all around, scanning for more tidbits. This one lacks any scarlet in its crown, a female, and also pays me no heed.

The female flies a few feet, and her frantic search reminds me of a saying my mother had. My mother used to say each time I didn't succeed at something, "Persistence and perseverance made a Bishop of his Reverence." I would stop and stare at her, not knowing what she meant at all but not about to ask, for fear of having another of her famous phrases descend upon me. Kinglets survive in spite of the cold winter, in spite of all the predators that could eat them, in spite of the other birds that are bigger and feed on the same winter foods. They simply keep on going, not letting the "Mr. Bobs" of the world get them down. Fifty years after my high school friends focused more on warblers, the demitasse bodies of kinglets and their pitcher-sized spirit still seize me. "No" is not an option for them.

My fingers clench tight into a fist and back out, again and again. The cold seeps through my mittens and socks. I flex up and down on my toes to push blood through my body, and yet I remain watching these miniature birds find morsel after morsel, moving quickly but methodically through this frost-covered patch like they are pulling things from the kitchen cupboards for dinner.

This is my home, too.

Chapter 15

In the Eyes of the Condor

Christina Baal

One dream, three raccoons, and countless blisters later, I finally find myself looking over the edge of Pacific Highway 1. It has taken a plane ticket that wiped out my checking account and a week of chasing down buses to finally get here, but I am overcome with a view that already makes my journey worth it. At the bottom of the 200-foot drop from the guardrail to the Pacific Ocean, waves pummel the cliffs. Wind buffets the scrubby bushes and hardy flowers that bloom in California in late February; I cannot fathom how they manage to cling to the edge of the earth. The sun strikes the ocean so fiercely that I cannot look past it at the infinite horizon. It is hot, I am sweating, my backpack scrapes into my shoulders, and cars are hurtling past me at sixty miles an hour. It is here, at the edge of the North American continent, that I am going to attempt to find the king of North American vultures, the bird I dream of finding above all others: the California Condor.

To anyone who has ever seen Disney's *The Lion King,* vultures are birds that fly around sinisterly waiting for Simba to die in the wilderness so they can eat him. But to birders like myself, vultures are these incredible creatures that clean up after us humans and keep the world free from a myriad of diseases. Although my family thinks that I am crazy, I insist that these bareheaded birds that pee on themselves to keep cool and vomit as a means of self-defense are the most beautiful birds in the world. Naturally, I have to find the most awesome of them all.

As a group, birders tend to skew towards crazy. While there are many birders who are content to watch the avian world unfold at a feeder from the comfort of an armchair, there are just as many who could never aspire to such sane bird-watching. I say this lovingly, because many of my favorite people are the kind of birders who spend more money on binoculars than they do on rent and call in sick to work in order to wake up at 4:00 a.m. to drive eight hours to stand for twelve hours surveying a six-square-foot patch of marsh looking for a duck that, to the average person, looks the same as every other duck. (In my defense, I was not the person who came up with this plan, and furthermore, a Garganey has a white crescent eye stripe and silvery wing feathers that make it quite distinct from your average Mallard.)

That being said, when I jump off the bus at the northernmost point of Big Sur and find myself standing alone on Pacific Highway 1, I cannot help but wonder if I am being too crazy even by birding standards. As a twenty-four-year-old wandering bird artist, I am not exactly in a financial position to buy my way out of trouble (or even rent a car, which would have made my weeklong logistical masterpiece of bus connections punctuated by miles of walking so much simpler). In case anyone was wondering, it is not easy to get from San Francisco to the northernmost point of the condor's range located in the Big Sur region without a rental car in the middle of February. It turns out that there is only one bus that will be returning from Big Sur to Monterey this week, and if I miss it tomorrow, it will be quite the feat to ensure I can reverse my trip to catch my return flight home to New York State.

So after hours of earning extra money to buy my plane ticket and hours of traveling to reach this one spot along Pacific Highway 1, I now have fourteen daylight hours to lay eyes on this bird I have wandered so far to find. Still, even this short amount of time to find a California Condor is a lot considering that only thirty years ago the condor seemed doomed to extinction. Although it had existed for over two and a half million years, evolving to survive the Ice Age and the postglacial world that followed, humans threatened the condor, so that even the possibility of finding one soaring above the skies in Big Sur is nothing short of a miracle. Now alone in its genus, the bird that became *Gymnogyps californianus* once existed in a Pleistocene world filled with mastodons, ground sloths, camels, and a slew of other giant mammals. It commanded a range that stretched from Canada to Mexico, from the West Coast along the Pacific Ocean to the East Coast as far north as New York. Scavenging condors made quick work of the gigantic carcasses that littered the land, perfectly adapted to a world that required the cleanup of creatures as enormous as they were. But the condor's world was destined to change for the worse. As the Pleistocene megafauna disappeared on the heels of the retreating Ice Age, the prehistoric birds found themselves surviving into a new era alone. The California Condor is a species that cannot adapt fast enough to an evolving human world. Today, the range of the California Condor is so reduced that there are very few places I could have gone to find it. There are wild populations along California's southern coast from Big Sur to Ventura County and in northern Baja California; there is also a small population in Arizona in the Grand Canyon.

A quick scan of the cliffs in the distance produces a Turkey Vulture, the smaller cousin of the California Condor. Even though it is far away, I can tell almost immediately that it is not the bird that I am looking for. Aside from the fact that I have logged hundreds of hours watching Turkey Vultures and know them better than any other bird, I can tell that this bird is too small to be a condor. The California Condor is the largest bird in North America, and its nearly ten-foot wingspan inspired fear, wonder, and legends amongst the native peoples of the Pacific Northwest; in modern times the huge bird has been mistaken for a small airplane. There are other raptors riding the thermals created by the cliffs that rise above the highway, and although I take the time to identify them all I already know that they are not condors. Something within me is sure that the moment I see the giant scavenger I will know it.

Condors have long been recognized as something out of the ordinary. Seventeenth-century Europeans exploring the Pacific Coast made sure to record their encounters with the enormous birds. In 1602, a Spanish ship charged with mapping the California coastline passed the carcass of a whale being eaten by condors. Antonio de la Ascension, a Spanish priest aboard the vessel, described these "birds the shape of turkeys" as "the largest I saw on this voyage."[1] Unfortunately, this is the extent of his report. My artist imagination fills in the details with images of the great birds soaring above the beached carrion, fighting for choice positions to gorge themselves, their red eyes flashing, their huge black wings flapping with eagerness as they plunge their pink, wrinkly, naked heads into the rotting flesh. I look absentmindedly over the guardrail that flanks my right side and can practically see the spectacle unfold through the haze on the spit of land beneath me. Consumed by my reverie, I lean too far over the guardrail and slip on the uneven ground beneath my feet. My heavy pack prevents me from regaining my balance, and I split my knee as it connects with the unkempt pavement.

Chasing rare and elusive birds often entails a certain amount of masochism. It is moments like this—when I have blood dripping down my knee—that a crack in my enthusiasm allows me to see how crazy I am. As the sun climbs above, I can feel sweat drenching the space between the pack I carry on my back and my already aching shoulders. I cannot tell how far I have walked, as the highway seems to go on forever. The curves of the road and the faces of the cliffs all blend together and look the same in the heat haze. Every once in a while a car whips around a bend, and I wonder how many vehicles rip around the bends too fast and slam into the same guardrails that had prevented me from hurtling over the edge of the cliffs as I slipped. I wonder what they think of me, the drivers who hurtle by and see a solitarily backpacker walking along the thin shoulder of this treacherous roadway. They have to think that I am ridiculous.

I have done a lot of reckless things while looking for birds, but it is usually only in hindsight that I identify them as such. Kayaking in a thunderstorm looking for Sora rails? No worries. Biking sixty miles in ninety-five-degree heat looking for Purple Martins? Not a problem. But perhaps it is because of the enormity of the bird I look for that this somehow feels like the most intense thing that I have ever done, and I am fully aware that my walk down Pacific Highway 1 is dangerous. With blood caked on my jeans, I suddenly feel, if only for a moment, vulnerable and alone in this beautiful landscape.

While my quest for the condor is my most ambitious adventure yet, it pales in comparison to some of the stunts other birders have pulled off in pursuit of birds. They will jump on planes and crisscross the world for a glimpse of a bird that finds itself out of its range, scale mountains covered in snow and ice, schlep through swamps riddled with venomous cottonmouths, and hitchhike across the country eating cat food to survive, all for the sake of watching birds. While ours is a way of life that might be confusing to someone who has not fallen under the spell of these winged creatures, it takes only a single moment to look at a bird and see the intricacy of its feathers, the awareness in its beady eyes, the wonder of evolution in its flight—and then, in that moment, to find oneself transformed by the sheer beauty of its existence.

I experienced one of those moments four years ago, and now something shifts within me whenever I go looking for birds. They instill a desire that feels all-consuming, that drives me towards adventures I never even dreamed of. And the condor, as the bird that owns my dreams above all others, is pushing me to greater lengths of crazy. Since arriving in San Francisco a week ago, I have pulled all kinds of stunts the likes of which I would never tell my mother. She does not need to know about the hollow I carved out to sleep in under a fallen log in Golden Gate State Park (although it was really quite cozy, and I was serenaded to sleep by a Great Horned Owl when the sound of traffic finally died down). There is also no need to hear about how I spent another night running after buses through the town of Gilford, dodging cars and racing through traffic jams before I realized that I had another chance to catch the bus in the morning.

My crowning misadventure, though, occurred midweek when I awoke at around 3:00 a.m. to a raccoon sitting right next to my head. It was eating out of my backpack, nonplussed when I sat up close enough to reach out and pet it. A little peeved, I tried to shoo it away, but it ignored me as it pulled the items out of my pack, one by one. To its credit, the bandit politely set my binoculars, maps, and socks in a small pile off to the side. But whenever it found something edible—I cringed as it pulled out my box of Trader Joe's chocolate biscotti, my big grocery splurge—it would bolt off a short distance away, devour it, and then come back to resume its sorting. Eventually I decided I might as well help it along, and while it was off stuffing itself I finished sorting all of my belongings. My pack now empty, I decided to take the opportunity to run to the restroom. On my return, I burst out laughing with a mix of exhaustion, exasperation, and hilarity. *Three more* raccoons were sitting on their haunches waiting for me. Needless to say, I did not get any more sleep that night.

Even though those raccoons ate all of my food, my pack still feels like I am carrying bricks as I continue walking along Pacific Highway 1. I wish I could just ditch it and come back for it later. I could walk so much faster without it, fighting against the clock that is rapidly ticking away my precious time.

As a bird that managed to evolve its way through two and a half million years of geologic and environmental changes, condors seem hardly threatened by time. But time almost did run out for the condor. No matter how much it adapted to survive into the modern era, the condor could not withstand the decades shaped by human influx. Inadequate food sources—after all, dead squirrels and rabbits are significantly smaller than the megafauna carcasses the condors were accustomed to—stressed condor populations even before westward expansion. The settling of California exacerbated their decline, by stealing much of their habitat. Egg hunters and scientists made the situation worse by collecting eggs and specimens. Adult birds were shot for sport and out of fear. In this time in environmental history, the magnitude of extinction had not yet taken hold of the public consciousness. Passenger Pigeons, Carolina Parakeets, and Eskimo Curlews had already disappeared, and the California Condor seemed like it could be next. But few people, if they knew what a condor was, even cared. More people considered the bird an obstacle in the way of progress, as laws to protect its habitat inhibited city development. By 1987, the California Condor had been reduced to a population of twenty-two.

Environmentalists went to war with each other over what to do to save the condors. A difficult choice had to be made, as it was clear that they were dying in the wild and would continue to do so if humans did not intervene. Despite protests, it was finally decided that all remaining condors would be captured and brought into captivity so that they could be kept safe and could breed to stabilize the population. This decision brought outrage from those who believed that this was the end of the wild condor. They argued that even if the birds could one day be released, they would be neutered versions of their predecessors. The last wild condor was a bird that biologists had nicknamed Igor. He fought hard for his freedom, eluding the traps set by scientists for months, until he was finally captured on April 19, 1987. On that day, the lineage of an ancient creature over two and a half million years old was broken. And with the capture of Igor, the fate of the California Condor was, in a dark twist, entirely in the hands of the same beings that had almost obliterated it.

Despite over thirty-five million dollars of conservation work and years of tireless recovery efforts, in the year 2016 it is still uncertain whether the hours that I have to find the condor are borrowed. The vulnerability that I feel as the cars hurtle past me is a fitting sensation while searching for a creature whose life is constantly at stake in a world ruled by the whims of humans. Today, impressive recovery efforts have restored the condor to a population of 446 birds. Four years after the condors were brought into captivity, the hard work of those devoted to the condor's recovery had increased the population from twenty-two to fifty-two. To my delight, I discovered that it was in 1992—the same year my life began—that the California Condor had fought back hard enough to merit the release of a handful of the captive-born birds into the wild. At first, the newly released condors horrified scientists, who watched the captive-bred condors approach humans, electrocute themselves on power lines, and die from eating antifreeze. But with aversion therapy, mentoring from once-wild condors, and improved efforts in hand-raising chicks, eventually, the released birds figured out how to survive. Twenty-four years since that first release, 276 birds fly free. While these numbers seem incredible compared to the population in 1987, they are still staggeringly fragile. It seems outrageous that I went to high school with more people than the population of an *entire species*. But regardless, the story of the condor is one of resilience, and it is this aspect of the condor that seduces me most.

There are many reasons birders chase birds. There are some people that chase them for the same reasons that other people collect rare coins or stamps, enjoying the rush of success after trying to find something elusive. Still others view finding birds as a competitive sport, trying to outdo other birders by seeing more birds than anyone else. I watch birds for many reasons. I love the rush of adventure and I love the stories I can tell after my trips. But even more importantly, for me, I look for birds so that I can paint them. And I paint them because I love them. It is a one-sided love for sure, a construed relationship between the romanticism of my human heart and the creatures that recognize my existence but for a few moments. But I love them nonetheless. I love the sense of wonder I feel as I watch them live out their lives, I love the way they make me realize there are more colors in the world than I could ever hope to paint, I love the cyclical nature of their lives and how my own life has adapted to mirror those rhythms. But most of all I love them because they make me feel alive in a way that nothing else can. I learned to dance with carefree abandon after watching Sandhill Cranes dancing in Wyoming; the Turkey Vultures of the Hudson Valley taught me balance as they alchemized the dead into new life by consuming roadkill. I watched the tenderness between Canada Geese and felt moved by their loyalty to one another; no sound in the world makes my heart leap like that of a Yellow Warbler in spring. And in this sense, I know that in searching for the California Condor I am hoping to find something that inspires beyond anything I have ever experienced.

Lost in the reverie of the condor's history and the physical exertion of my walk, I am taken aback to realize that I have been walking for over six hours. The sun is no longer directly overhead, but it is still fierce. There is a small copse of bushes that rises up over the guardrail a few hundred yards ahead of me, and I head towards it with the intent to take a short break. To my surprise, as I draw closer I notice that there is already someone availing himself of the shade. As I draw level with him, he looks up and waves merrily before addressing me. "Are you looking for the condors?" My heart leaps. This person is suddenly the most important person in the whole world. "Yes!" I answer him with so much enthusiasm that I pump my arms in the air. He laughs and gestures at the binoculars around my neck. "This isn't exactly the most social highway in the world. I don't see that many walkers. Anyone crazy enough to be walking along with binoculars can only be looking for one thing." He continues, "There is a small cottage about a mile away. There is a tree nearby. The condors sometimes sit there when the wind dies down. It's been pretty windy today, so I don't know what the chances are that they'll be there, but it's a good place to try to see them." I could have hugged this man, but I restrained myself, worried that my enthusiasm might knock him backwards into the bushes. Thanking him, I suddenly feel as though I have the energy to sprint up the road despite the heat and the weight of my pack. I have a chance. I repeat his words over and over, feeling so excited I am shaking. Sure, a chance is all that I ever had, but now it seems real. My heart is racing. There are really condors here. *I could actually find one.*

My whole body surges with adrenaline. All of the effort I have put into this quest feels like it has amounted to this moment of possibility. I scorch my eyes by keeping them glued to the sunny sky, and I jolt at everything that moves. I am so close. I know that it is still a long shot, but I am giddy with excitement. While I could not have undertaken such a masochistic quest without some semblance of a belief that I could be successful, at the same time, I had never allowed myself to forget that the chance of failure was infinitely greater than the chance of success. But now I could no longer hold myself back from believing that it was really going to happen. Finally, I let myself feel real hope.

Even as the population of California Condors continued to grow, "real hope" eluded the members of the Condor Recovery Program for years as they encountered setback after setback. Aside from the captive-bred condors that had trouble acclimating to the wild, scientists had to deal with constant upheaval over the reintroduction of condors to land that was coveted by developers. And besides all of this, scientists had finally recognized the threat of lead poisoning. Wild condors were at constant risk, as there was no way to ensure that a healthy bird did not consume a poisoned carcass and accumulate the lead toxins in its body. Poisoned condors can be saved if they are found and treated in time. Despite stringent laws to protect them, condors are still at risk of being poached. After surviving lead poisoning, the forty-year-old once-wild bird known as the Matriarch was shot and killed by a hunter who claimed he did not know that the bird was an endangered condor.

Still, despite such blows, there were moments to celebrate. A separate population of condors established in Arizona in 1996 ensures that a single disaster cannot eradicate the entire population of wild condors in one blow. In 2003, the first nestling fledged in the wild since 1981. In 2015, more condors were born in the wild than died. And, I cannot help but feel that poetically, although interrupted, the lineage of the ancient California Condor was resumed when on February 4, 2002, the process of restoring Igor to the wild began on the same day I turned ten. Fourteen years later, he is still alive and well, soaring above the Hopper Mountain National Wildlife Refuge in Ventura County.

And it is with Igor and his freedom in mind that I turn around the bend and time stands still.

I feel as though I am in a vacuum in which the rest of the world does not exist. My body seems weightless as I all but run up the hill towards the outcropping where the two dark shapes rest in the single tree that juts out just above the cliff face. I reach the guardrail and clutch it with shaking arms. Cars that pass by feel as though they are in another universe. Two condors. There are two California Condors in that tree. Their bare pink heads and necks seem so soft against the striking contrast of their black-and-white wings that stand out sharply despite the heat haze. Even from a distance, it is clear the condors are enormous; they are visibly larger than any raptor I have ever seen. And no matter how ridiculous I know it is, I indulge the feeling that they are sitting there waiting for me.

The California Condor was known as the "thunderbird" to many tribes of the Pacific Northwest. It was believed that its wings were so powerful that each wingbeat summoned thunder; the Tlingit claimed that lightning flashed from the bird's red eyes. The Gashowu Yokuts tribe of south-central California told stories of a condor whose wings caused eclipses. I devoured these stories before my search, allowing my mind to create the image of a mythological being whose very shadow instilled fear. But as I stand gaping beside the guardrail, the roaring of the car engines is suddenly muffled by what sounds like a cracking boom. Something primal takes over my body and forces me to my knees. A mixture of fear and wonder the likes of which I have never known washes over me. *The stories are true.* I recognize that this line of thought is absurd even as I am engulfed by the enormous shadow of a third condor. I look up to see the thunderbird right above me, its huge black wings carving a swathe of darkness in the otherwise bright sky. The diagnostic white feathers in the center of each wing are visible. In my delirium I cannot tell if the California Condor is a behemoth or if I have shrunk within the force of its presence. Never before have I realized that a shadow can feel so heavy. The condor beats its wings once more as it soars overhead, so close that I can see the individual feathers in its wing coverts. *There really is thunder in its wingbeats.* Even though I am aware that I have become lost somewhere between a waking dream and reality, I know that the force of that wingbeat is real as I feel it thrum through me. I half expect the sky to darken in the bird's wake. But then the shadow passes as quickly as it appeared, and I whip around to watch the condor soar above and beyond the cliff face and disappear.

I pick myself up and lean over the guardrail. In all of my wildest condor fantasies, this was more than I ever could have hoped for. *Three condors.* I cling to the image of what just occurred, the force of the shadow still upon me, the crash of thunder still echoing in my ears. For the first time in days I feel calm. I drop my pack to the ground and lean against it, stretching out one leg and bending the other so I can prop up the arm that holds the binoculars glued to my face. And I watch them. I found the condors, and now nothing else in the world matters.

These condors in the tree radiate serenity. Unperturbed by the wind, they set their gaze in the direction of the flickering ocean. Their movements are slow and few. For long stretches of time they do not move at all. Only the feathers of the thick black ruffs around their necks dance in the wind as though to prove that the condors are made of flesh and bone and not the stone of the cliffs below them. They are nothing short of majestic, these kings amongst vultures, as they sit atop their throne.

I do not know how long I watch the condors before they begin to stir. The one on the left stretches its dark wings and flaps them gently, turning as it does to face the other bird. It dips its head and rests it against the neck of the other condor before beginning to preen the ruff of its companion. I marvel at how gentle its movements are. The bird being preened closes its eyes, looking quite content, before lifting its head so that it can reciprocate. Something about the tenderness of the exchange makes the two birds feel extraordinarily vulnerable. I wonder if they are perhaps a mated pair, bonded together for the rest of their precarious lives. No matter how awing they are in flight, by conventional standards the California Condor is an ugly bird. Like all vultures, its head is featherless and wrinkly. Lumps of white and gray bacteria often accumulate on its face. A flap of skin beneath its chin makes its skin seem stretched and decrepit. But everything about these two condors is absolutely beautiful. As I watch the two condors preening each other, so alive in their intimacy it hurts, I feel deep shame at the cruelty of the species I am part of. What right did we ever have to gamble with their existence at all, to deem them unworthy of fighting to survive, to push them toward oblivion because we decided they were ugly, or a nuisance, or had no value in this world?

The California Condor has long been a symbol of the environmentalist movement in the United States. Its comeback is a beacon of hope for the future of environmental conservation. But it is also a warning of the fragile relationship between humans and the natural world. The condor escaped extinction, but over 1,500 of the world's 10,000 species of birds are endangered. Over 5 billion birds die every year in the United States, meaning that over 13.7 million birds die every day. Already, I will never get to see a Passenger Pigeon, and I almost was denied the chance to lay eyes upon a condor. My heart sinks, wondering what birds will vanish in my lifetime.

I realize that I am no longer alone as I watch the condors. Cars have stopped at the nearest pullout, and their drivers have gotten out, intrigued by the unusual sight. It occurs to me that as none of them have binoculars they are not likely birders, yet they still recognize that the enormous birds are something special. A mother lets her two children sit on the guardrail next to me. Although she keeps her hands tightly around their waists, I laugh aloud at this rather dangerous move to help them see the condors—if they fall, they will plunge onto the sharp rocks below the cliffs—but I laugh because really, who am I to judge? They look at me as I chuckle, and I pull my binoculars over my head and hand them to the child beside me. He looks like he is about five years old. I grin, wondering if this will set him on a path to becoming a crazy birder as his eyes widen in wonder, my binoculars bringing the wrinkly folds of the condor's head and its jet-black feathers into sharp detail.

I have become so interested in watching the children as they pass my binoculars back and forth that I have not noticed the wind pick up. A sudden movement turns my attention back to the condors. They are no longer preening; instead, one of the birds has begun to stretch and flap his wings more emphatically than before. The other bird gets up and moves around in the tree. And then, before I can realize what is happening, the first condor launches from its perch, wings stretched wide, and lets the wind carry it smoothly towards the guardrail. Its powerful muscles pump its enormous wings, and the notches in its worn primary feathers look longer than my forearm. I cannot breathe as it draws level with us, so close that I can see the small hairs on its wrinkly head even without my binoculars. As it passes me, my brown eyes meet its fiery red ones, and all at once I am drowning in crimson.

There is an argument over whether or not the radio-tagged, highly managed condors, even when released, can be considered truly "wild." But definitions mean nothing to me as I look into the eyes of the California Condor. *Defiance*. There is no other word for the force that radiates from the wild condor's eyes. I recognize that I am looking into the soul of a being that has challenged fate and won, that is invincible in its eternal mark on the world. For better or for worse, the lives of condors and humans are intertwined. And now, by looking into its eyes, I too have become bound to the California Condor. Whoever I was before I fell into those eyes, I am not the same when I resurface.

Gasping for air, I watch the first condor as it rises up, up above the cliffs that stand above the pavement. The second condor follows it before I can catch my breath. The two of them circle a single peak, taking turns eclipsing the sun with each pass. Their sky-dance is one that rejoices in their freedom as their great wings carry them triumphantly through the Californian winds that are their birthright.

My heart is surging with hope as the condors climb higher and higher, continuing to circle the peak. Delirious in my joy, I care not that I have done something as naive as ascribing the fate of the world to a bird. I could not have known then, as my twenty-four-year-old self watched the condors disappear, just how fiercely the king of the vultures would continue to inspire as I wander the world searching for the creatures I love. But I did know that for as long as I lived I would remember the feeling of staring into the eyes of the California Condor. No matter how crazy I may be, I would not trade my wild life for any other.

I look away from the condors for just a moment, tears forming at the corners of my eyes. At least fifty people have gathered along Pacific Highway 1 to watch the California Condors. They are all sizes, colors, and ages, coming and going from who knows where. But none of that matters as they gaze skyward, transfixed by the sight of the ancient creatures that have traveled through space and time to bring us this moment. And it is in this beautiful instance that I realize that the birds we almost destroyed are, in fact, the very things that could save us. Humans might make horrible decisions—we might be terrible to the natural world, and to each other—but, like the condors, perhaps we have a chance. At the last moment, we recognized that we needed to save the condor, perhaps giving us a chance at some sort of redemption. The Wiyot tribe tells a story of how the condor saved the world from a flood meant to wipe out the human race. Perhaps, one day, if the world we are breaking is poised to fall, red eyes of defiance will stare into our souls and save us all.

Chapter 16

Little Brown Birds

Richard Bohannon

There were two gates when I first drove up. A wildlife biologist at the refuge had directed me to where a bird had been seen down a small path, just two ruts indented into the prairie. I didn't see anything at the first gate, but the second led up a slight incline to a chain-link fence. She'd instructed me not to block the gate in case maintenance trucks needed to enter, so I pulled off far to the side.

The morning was chilly for early June. I grabbed my jacket and loaded it up with a granola bar, a small water bottle, and my guidebook, assuming the path ran out from somewhere near the fence and onto refuge land. It didn't, as I saw after walking just a few steps up the drive. The outer gate where I had parked was open, but led only to a nondescript gravel lot, clearly fenced off and leading nowhere; the biologist had mentioned a silo, so I figured this must be some abandoned military site. With nowhere else to go, I pulled out my phone to play a Baird's Sparrow call—I knew the bird had been at least seen in this general area, and perhaps I'd be able to hear one call back from this little gravel driveway. The expected silence followed, and I walked back to the car.

This was the fourth and final day of a trip to western North Dakota's Bakken region, where I was mapping out habitat fragmentation caused by the recent surge in oil development. My research was really just an excuse to go out birding for a few days—all of the actual work involved analyzing aerial imagery back home on my laptop, but I was trying to correlate the habitat loss to the decline in bird populations.

Two birds are confined to the northern mixed-grass prairie: the Baird's Sparrow and the Sprague's Pipit. Both are small, brown birds, not terribly charismatic—what birders call LBJs or "little brown jobs"—and both are declining in population. I had been out to the Bakken during the previous two summers to do some light research and birding, but had failed to see either bird; I organized this trip specifically to see them.

The first three days had come up short for both birds, so I'd set a 5:00 a.m. alarm for this final outing. I was aiming for Lostwood National Wildlife Refuge, about an hour away from the cabin I had rented.

Before the recent oil boom there were few people in this part of the country—as recently as 2006, some towns would give land away for free to those willing to relocate there,[1] and even today it's only thinly settled. When driving along roads with oil development, large extended cab pickup trucks and semis regularly sped past my little hatchback, sometimes in large-enough numbers to jam up the few large paved roads. Away from the oil I was almost totally alone, though, hours passing without seeing another person.

Oil development occurs with seeming abandon in the Bakken and is only lightly regulated.[2] Despite a history of progressivism and socialism in the American prairies, North Dakota today is essentially a one-party state—an explicitly oil-friendly Republican Party has held the governorship and both houses of the state legislature for years. Unlike resistance in parts of the East Coast (fracking is banned in Vermont, New York, and Maryland, for instance, as well as in several cities across the U.S.), there have been no large-scale protests in North Dakota, save resistance to the pipeline by the Standing Rock reservation.

The Bakken region is named after its oil-bearing geologic formation, lying about two miles below the surface. Geologists and oil companies have known about it for decades, and the region had an earlier, smaller boom in the 1970s. But the Bakken's oil is harder to extract than conventional oil, and this first boom was short-lived; the oil here rests in a very thin layer, averaging only about thirty feet in depth, and it's infused within shale rock, making it "tight" oil that doesn't flow freely.

Despite its newfound notoriety in the public consciousness, fracking has been around for a long time. Its first commercial use came in 1949, courtesy of Halliburton, and the concept itself dates back as far as the 1860s, when nitroglycerin was used to stimulate the initial flow of wells.[3] By the 1980s it was a common means of extending the life of existing conventional, vertical wells. Newer forms of fracking combine it with horizontal drilling, with the oil well turning ninety degrees and traveling horizontally for a mile or two within the rock formation.

While the length of these horizontal wells isn't visible at the surface, the oil industry is omnipresent. The northern unit of Theodore Roosevelt National Park, for instance, lies around and within canyons carved by the Little Missouri River. It's a solid hour from the interstate and has no hotels, cabins, or even showers, so it's not too hard to find yourself alone there on a trail. And yet distant well pads are visible from many of the park's higher spots, and occasional loud, echoing booms can be heard in even the most secluded canyon bottoms, emanating from fracking equipment.

Large well pads are a common sight when driving through the region to Lostwood, often along newly built and dusty gravel roads. Large flares dot the night sky, burning off the excess natural gas; development has been too rapid for natural gas infrastructure to keep pace, and the gas is often simply burnt off while the more valuable oil is extracted. Towns are expanding to meet the influx of oil workers, and rail lines and pipelines are expanding to move the crude oil out to refineries; these lines extend the hazards of spills out into the rest of the state.

Lostwood itself is ringed by several well pads immediately outside the refuge boundaries—there are about a dozen just over a mile from where I pulled off in search of the Baird's Sparrow, and the ubiquitous oil development, not the more subtle imprint of nuclear warheads, marked my mental map of the area that morning.

Having come up empty, with neither an interesting bird nor a path to find one on, I moved my car back to the first driveway, initially just to stand there and listen for a few minutes before driving away. Only after walking directly up to the small gate did I notice that the tall grass hid two ruts in the ground, marking the trail I'd been seeking.

The hike started out more quietly than usual. A few Red-winged Blackbirds called from a patch of wetland by the road, and a few Bobolinks gave their bizarre jumble of a song. Eventually the same trinity I'd been hearing for the last three days slowly emerged—Grasshopper, Savannah, and Clay-colored Sparrows.

After a slow half mile, with lots of stops to listen and scan with my binoculars, my heart rate jumped as I started hearing something new, something that seemed like the end of the Baird's Sparrow call. The typical call has three soft pecking sounds, followed by a more melodic trilling whistle, and I could hear only what might be the whistle; whatever bird it emanated from was still a good distance away. Perhaps it came from my elusive sparrow, but like a lot of birders I'm hesitant to ID something until all other options are discounted.

Something small flew past the corner of my vision, and I turned around to track it. It was just another Savannah Sparrow, but following it had me looking back toward my car, which was no longer alone.

A large white truck had parked behind it, both blocking the gate and boxing me in. Too far away for me to make out any details, it looked like a delivery truck or a large pickup with an oversized enclosure covering the bed. The product of an evangelical upbringing, I immediately felt guilty and wondered if I had inadvertently blocked the gate and was now in trouble with the refuge staff. My binoculars told a different story, though: a man in military fatigues was looking through each of my car windows, thoroughly and methodically, and speaking into a radio.

Getting in trouble with the staff at a wildlife refuge is not something to be proud of, but it's otherwise a relatively benign offense. Having your seemingly abandoned car searched by the military raises the stakes, however. I immediately started walking back to the car, but I was at least a half mile away, wearing green and surrounded by prairie grass. It was much too far for him to hear me call, and too far for him to notice me. Waving at him wasn't doing any good, but I kept at it; I didn't want to run, as it might seem aggressive, and he had a large gun. My main concern at this point, given his interest in my car, was that the unmarked white truck was part of a bomb squad, and my car would soon be exploded (or, less dramatically, towed away). In retrospect this probably was never a real possibility (at least the exploding car part), but it was clear to me that my trouble here stemmed from the possible-nuclear-missile silo next door, and not from blocking an underused fence.

After some indeterminate time—it felt like quite a while, but it was likely less than a minute—another truck pulled up, this one unmistakably military. Out of a tan, armored Humvee came three new men in camouflage, machine guns pointing out and scanning the silo site. Another soldier sat atop the Hummer in its armed turret.

My hand waving took on a new urgency at this point. Still unseen, I hurried toward them with both hands up, my right one waving, a pair of binoculars inelegantly swinging around my neck as I clomped through the prairie grass, anxious about the armed men ahead but also about whether I'd be able to find the whistling bird again.

Nuclear missiles came to the North Dakota prairies as a child of the arms race following World War II. After the United States dropped two atomic bombs on Japan, ending the war, the Soviet Union was quick to finish developing their own weapons as a deterrent. The U.S. followed suit by developing even more, and more powerful, weapons, and the arsenal of both countries quickly grew into a malevolent absurdity.[4]

While eight other countries also have nuclear bombs, these weapons were and continue to be overwhelmingly concentrated in the U.S. and former Soviet Union. The total number of weapons maintained by the U.S. peaked in the 1960s at over 31,000 warheads, but the U.S.S.R. continued to accumulate them well into the mid-1980s. At the global peak in 1986, there were over 64,000 nuclear weapons across the world. Numbers have since declined with the end of the Cold War; globally there were over 9,200 in military stockpiles in 2017, with the U.S. and Russia still holding over 4,000 each.[5]

Roughly underneath my feet, while I had stood dumbly playing a sparrow's song, had been one of these bombs: a Minuteman III missile carrying a thermonuclear warhead many times stronger than the bombs dropped on Hiroshima and Nagasaki, and capable of hitting a target on the other side of the world in only thirty minutes. This is one of 150 buried across North Dakota, roughly a third of which are in the Bakken. First deployed in 1970, there were initially 1,000 of them spread across the plains, along with other intercontinental ballistic missiles. As of 2017 only the Minuteman III is still in commission in the U.S., with 450 buried in rural areas of Wyoming, Montana, Nebraska, Colorado, and North Dakota. Some were previously equipped with multiple warheads on each missile; today each contains one bomb.

Silos embody a potential violence not only against the enemies of the United States government but also for the people and land where they're buried, either from an accident or as the target of an attack. Eric Schlosser has documented the 1980 disaster at a Titan II nuclear missile silo in Damascus, Arkansas, for instance: an accidentally dropped maintenance tool damaged the missile, causing it to slowly build pressure and eventually explode, injuring several people and killing one soldier.[6] The warhead itself thankfully did not detonate; if it had, millions could have died.

I only came to know all of this after the fact, in the safety of a library. At the time, the missile silo's nondescript and fenced-off square of concrete had simply been a reference point in my search for that two-rut track out into the refuge.

After waving and tromping along the trail for another minute or two, I saw the soldier closest to me—still out of shouting distance but now close enough to see clearly without binoculars—wave back. He and the other two soldiers scanning the area began to slowly return toward my car. Keeping my hands up, I walked the next couple of minutes in awkward silence and stopped at the little red gate.

"Hop back over," the soldier in the turret directed me. He wasn't overly aggressive, but he didn't need to be: they had an armored Humvee and multiple machine guns, each held with both hands, though not pointing at me.

"What are you doing out here?" the same soldier asked after I hopped the gate. He seemed to be the one in charge of this unit.

"Out, ah, looking for birds."

"You're what?" He seemed genuinely to not hear me, and not just finding my answer absurd.

"I was bird-watching," my answer a bit louder this time. It was quite windy, so the sound of rustling prairie grasses created some background noise.

"Bird-watching? Can we see some ID?"

"Yeah. I'm going to reach into my back pocket. Is that okay?"

"That's fine, and you can put your hands down," he replied.

I passed over my driver's license. The soldier closest to me asked if I had any secondary ID, so I passed over my university ID card as well.

"So you're a student?"

"No, I'm a professor," I explained, telling them I was looking for birds related to my research.

After they radioed in my ID and talked among themselves, I got another question from the turret: "Were you taking pictures of the site over there?" Throughout this whole exchange, he and the others always and only referred to the missile silo as simply "the site," which in retrospect has a sci-fi ring to it—it is that which cannot be spoken of directly, the thermonuclear elephant in the room.

"No, I wasn't taking pictures. I'm just out bird-watching. Someone at the refuge had given me permission to be out here."

The soldier in the turret paused for a second, then called back, with an air of mild skepticism: "So you don't have a camera with you?"

"No. I mean, yes, I have one on my phone but I wasn't using it." It occurred to me I had likely been filmed this whole time, and they'd seen me holding up my phone.

"I did have my phone out, but I was playing a bird call," I followed quickly, not wanting to seem a liar.

"You were playing a bird call?"

"Yeah. I have an app on my phone."

"Can you show us your phone?"

"Sure I'm going to reach into my pocket again."

"Yeah, that's fine," the soldier in the turret replied, not seeming interested in my caution. It's worth saying here that I'm a middle-aged white man driving a relatively new car, and all five of the military personnel appeared to be white men, young enough to be my students. If any racial profiling was going on, it was to my benefit.

I kept my hands visible during this whole exchange, just in case, but got my phone out of my pocket. The soldier closest to me came over to look at it as I turned it on.

Thankfully, and not surprisingly, my bird app immediately appeared with the Baird's Sparrow vocalization page up—so now all five of these young men have had the chance to learn this semi-rare sparrow's call. "So you didn't take any pictures of the site?" the soldier in the turret asked again, after I turned the phone around to show him my screen.

"No," I replied. The conversation went on like this for a while, me showing them pictures on my phone and explaining my earlier conversation with refuge staff.

Eventually they seemed satisfied I wasn't a threat. One of the soldiers had moved back into the armored Humvee, talking periodically to someone on the radio, with the soldier in the turret leaning down and giving more instructions.

"Give us a few more minutes to get a clear to let you go," said the one in the turret. "Sorry again for keeping you."

"Can I go back out there?" I asked. They paused at this, apparently not expecting it.

"Um, well, we don't have jurisdiction over this property, but just over the site," replied one soldier, who kept reminding me of a recent student.

The soldier in the turret added: "Let me call this in and make sure it isn't a problem."

This took a few more minutes, but eventually they got the clear from whomever they answered to. With a reminder not to take any pictures of the site—the reason for which I'm not entirely clear, as I've since learned the Internet is full of such images, including Google street view, and you can even download silo coordinates off Wikipedia—and a few quick goodbyes, I hopped back over the fence and out into the prairie, making sure not to turn around. A possible Baird's Sparrow awaited me, and I was going to find it.

When I told my sister about all this a few days later she was surprised I had gone back to look for the bird. This, in turn, surprised me: not returning had never occurred to me. I had been concerned they wouldn't *let* me go back out there, but there was no question that I wanted to. Once it was clear the soldiers weren't going to arrest me, my concern had become whether or not the bird had flown off. She caused me to reflect, though: Why all this effort to see a little brown bird?

The Baird's Sparrow is somewhat rare, and it would be my first time seeing one. It's not listed on any endangered species list, but its numbers are dropping, especially in the United States (it also breeds in the Canadian prairies, where conservation efforts have helped stabilize the population). But while both of these elements add some intrigue, I'm mostly a lazy birder who likes to see the birds native to wherever I already am; I'm usually too cheap and too guilt-ridden about my carbon footprint to go on long birding excursions.

And while the refuge is lovely, the region is pockmarked with drilling rigs, gas flares, and large, idling trucks. Oil development is harsh and ugly. Well pads dot the land, roads expand to meet the new traffic, massive trucks pass you in a hurry, small towns inelegantly spill over in a manner akin to unpleasantly biological metaphors, and the small cities like Williston reek of a kind of masculinity I was glad to leave behind in high school locker rooms. This all fits in with how North Dakota, perhaps the quintessential flyover country, is often portrayed in the media. As a whole it receives little national news; when it does, the coverage tends toward the negative. Small cities flood (Grand Forks in 1997; Minot, and almost Fargo, in 2011), and the western part of the state was depressed and depopulating until oil came back and made it wealthier but polluted and corrupt.

But this negative vision misses as much as it shows. Violence and pollution don't tell the whole story of the Bakken, an open land where you can watch an entire storm system coming an hour ahead of time only to have it skirt just by you without raining a drop. It is a land of dramatic canyons carved out by the Little Missouri River, with Golden Eagles nesting on multicolored bluffs and orioles flying in the cottonwoods along the river. It is a place with vast prairies and gentle hills, where away from the well pads you can find yourself completely alone for hours—just you, a gravel road, and a whole lot of grass, without another person, building, or car in sight.

In this confusing landscape, maddening and ugly yet somehow also expansive and compelling, I think there was another reason pulling me to go back out there, beyond adding a bird to my life list or finishing my research: a little glimpse at something unique and beautiful, even if understatedly so. The Baird's Sparrow and the Sprague's Pipit, two unassuming little brown birds, embody this fragile landscape for me. They are all vanishing but are not yet fully erased, with a beauty too easily overlooked.

How do you grasp the beauty of a place while also mourning the violence we've imposed on it? Or can you mourn a place only if you've also been drawn to it, felt a charismatic pull? Our willingness to flirt with environmental catastrophe and nuclear violence on an unprecedented scale are laid bare in the Bakken, a place where we've also largely neglected to find beauty and value. We plant it with oil wells, pipelines, and nuclear missiles because, for most Americans, it's simply a place in between, a place without trees or interesting cities or even many paved roads.

The plains of North Dakota are a little brown bird, ignored and declining but still singing for those who listen.

I knew what the Baird's Sparrow looked and sounded like in my books and app, but birds themselves rarely use field guides and don't know to act according to our expectations. When I had heard the soft trilling earlier, right before noticing a camouflaged man peering into my car windows, I thought, "That might be it!" but with an emphasis on *might*. It could have been another bird I don't know the call for, or an odd-sounding Grasshopper Sparrow from a distance.

After leaving the soldiers, I walked more quickly to where I had encountered the trilling. My speed came partly from a reasonable desire to put distance between myself and their machine guns, but mostly I wanted to get back before my mystery bird flew away or quieted down for the day.

There aren't many obvious landmarks on the open prairie, but Lostwood is covered in gentle, low hills. I came across the area I remembered, where the path turned slightly to the north with a very low rise to the east, and waited. The trilling reemerged, but with the same problem as before—I could hear it intermittently, but not any preceding chirps, and it came from somewhere on the other side of the rise.

I generally avoid walking directly into prairie grass in the summer for the same reason I don't bathe in wetlands full of leeches: I know I'll get more ticks than birds. The path itself was already covered in tall grass, though, and I had expended too much effort at this point to be concerned about a few wood ticks, so I headed off through the grasses, stopping periodically to listen.

And there it was: after fifteen or twenty minutes, a single small sparrow emerged in the middle of forty-three square miles of open prairie, periodically flying up onto grass stalks to give a call before hiding again.

Chapter 17

The Keepers of the Ghost Bird

Jenn Dean

From the air, Bermuda resembles a jeweled and pregnant seahorse, hanging by its tail from the Sargasso Sea. Its top and bottom wrap around two ancient volcanic calderas, one at Castle Harbor and one at Dockyards. In between, Bermuda's twenty-one square miles twine along a fragile, curving limestone spine. From a taxi van's window, the coastline uncurls: the rose sand, the bitter green foliage of tamarisk trees, stunted windblown evergreens, and the blue southern tip of the Sargasso Sea.

Beneath the water, pillow-sized blue parrotfish gnaw algae off the reefs with their beaks, excreting great clouds of sand. It is said that all of the beaches on Bermuda have been through a parrotfish at least twice. When you walk anywhere, you are strolling on the tiny skeletons of reef creatures, eaten and recycled into sand, compacted over centuries into limestone. All of the limestone on Bermuda looks carved, as if someone took a giant comb and raked lines across the cliffs, revealing not the effects of jackhammers, but of ancient winds. Everything bends to the will of the wind: the birds, facing it with wings arched, stay in place as if pinned to the sky. The palm fronds rattle like bones, as if all the sailors' skeletons who died over the centuries on the infamous reefs have sprung to life to dance.

Depending on your view, Bermuda is either haunted or enchanted. The reefs lie beneath the waves, mysterious and invisible. There are rim reefs, patch reefs, champagne breakers, and boiler reefs. They extend so far north, ancient sailors always believed they'd sailed far enough around them, only to crash, even as the lights of St. Elmo danced on their masts for good luck. Glass ampoules of morphine from old wrecks are common souvenirs around the island; Bermudians used to induce ships to wreck upon the reefs for such bounty. It is written that you could see the flickering lights of people taking treasure off the boats. In one account, a minister interrupted his evening sermon, warning his parishioners to wait for him to take off his cloth so everyone had a fair shake at the treasure.

Bermuda was born of volcanoes and rumors. In the 1500s, the king of Portugal thought the cedar trees caused the winds and the ill weather; early sailors thought the abundant spiders indicated the presence of gold. Everyone agreed the islands were haunted. Unearthly cries came from the olivewood and palmetto groves at certain times of the year. Because of this, maps from the fifteenth century designate Bermuda the Isle of Devils, long before the Bermuda Triangle's birth.

In 1610, two people lived on Bermuda's one hundred and twenty interconnected islands. Four centuries later, sixty thousand people live in Bermuda, a "suburb in the sea." Enormous cruise ships pull into Hamilton—the only real city—every day at noon, negotiating their bulk through Dundonald Channel like fat men wading among the rocks. Once docked, they disgorge tourists who scurry the main thoroughfare of Front Street and disappear like hungry voles into the gay, expensive shops: once Trimminghams, now Blucks, now Smiths; shops selling fine china, silk and wool clothing.

Pastel houses dot the hillsides like confetti. Nearly every square foot of ground is developed. Roofs made of dazzling white steps, built to lure rain into cisterns, poke through foliage curtains of imported Surinam cherry, hibiscus, and Royal Palm. There are no streams or lakes. There are no woods for kids to get lost in. Schoolchildren walk the sandy lanes home, chanting and chaotic, wearing starched white uniforms.

The only wilderness left in Bermuda is the sea.

Though explorers from the fifteenth and sixteenth centuries were a sturdy, brave lot, Bermuda—a place of mystery and the bizarre—scared the hell out of them. Stuck as it was in the middle of the Atlantic Ocean, on parallel with South Carolina, surrounded by unpredictable weather and acres of invisible reefs, no one thought Bermuda a paradise. If a ship calculated its course wrong, or got blown off route during a storm as it headed from the New World back to Europe on the Gulf Stream, it often wrecked on Bermuda.

As if the hidden reefs, the awful winds, and the wild tempests around the islands weren't enough, devils lived on the islands, and would come caterwauling out at night during stormy weather, flying around the masts, crying "*diselo*" or "tell 'em"—as in "tell them we're from hell." Reluctant to shelter there during storms, most mariners believed Bermuda—named after Juan de Bermudez, who found it while searching for the fountain of youth in the early 1500s—was a pit of hell in the mid-Atlantic.

Bermuda was colonized by accident during the famous wreck of the Sea Venture in 1609, upon which Shakespeare based *The Tempest*. Bound for the new colony of Virginia from England, the Sea Venture separated from its convoy during a particularly strong blow, with Admiral George Somers on board. The storm grew into a hurricane and chewed on the ship for four days, spitting it out on Bermuda.

Miraculously, no one died. After Somers claimed the islands in the name of England, the survivors found not hell, but paradise. Abundant palmetto berries, olives, mulberries, and prickly pear cactus grew everywhere. The seas teemed with so many fish the sailors feared putting their toes in the water, and one could walk across the bays on the backs of the green sea turtles. A sturdy cedar forest furred the land, and they used the timber to build two ships, the Deliverance and the Patience, which would ferry most of the people off the island nine months later, leaving a few behind as the first colonists. As for the infamous devils, they encountered none at first; but one late fall night blackened by storms, the demons arrived. They swooped over the water, caterwauled over their heads.

How relieved, and ridiculous, Somers's men must have felt when they finally saw the creatures in their torch lights. William Strachey, a poet and failed theatrical entrepreneur who had been hoping to become employed as the Secretary at the Jamestown colony, described them in 1610:

> A kinde of webbe-footed Fowle there is, of the bignesse of an English greene Plover, or Sea-Meawe, which all the Summer wee saw not, and in the darkest nights of November and December (for in the night they onely feed) they would come forth . . . and hovering in the ayre . . . made a strange hollow and harsh howling . . .[1]

Over successive nights, the sailors discovered if they made outlandish cries, the birds landed on their outstretched arms. The birds, in breeding season, were locating their mates by sound, and curious, had never seen people. The men weighed them in their hands, and strangled or clubbed to death the fattest and heaviest ones for food. The sailors later discovered thousands of eggs in the devil birds' underground burrows. Within a few hours they gathered enough to serve one hundred and fifty men.

Prior to 1600, it's estimated that half a million pairs of devil birds bred on Bermuda, making it, in essence, a gigantic seabird colony. The cedar trees that covered Bermuda were endemic and low-growing; they tilted in high winds, uprooting and leaving small cavities beneath. The birds used their black beaks, which ended in a graceful hook, to dig twelve-foot burrows beneath the trees, and used their webbed feet to push the dirt out behind them. They lined the depression with feathers or grass, and laid a single egg. The parents took turns incubating for forty days, and after the chick hatched and dried, the adults climbed up the cedars using their hooked beaks and their feet, which had tiny nails on the ends of each webbed toe, and took off into the wind. Strong and intrepid flyers, the parents made dozens of round-trips from their rich Gulf Stream feeding grounds, where shrimp and squid came to the surface at night, then flew back to Bermuda to feed their chicks. They did this, nonstop, for months.

The sailors called it the cahow after its sound. It would be centuries before it would emerge as a species of gadfly petrel—a sleek-bodied, hollow-boned soarer with three-foot-long, paddle-shaped wings—but by then, it would be too late.

The cahow, or Bermuda Petrel, had no mammalian predators when the first Spanish galleons shipwrecked off the rock-strewn reefs sometime in the 1500s. The pigs the sailors left, however, likely ate several hundred thousand cahows before the Sea Venture even left England, and in the decade following the Sea Venture's arrival, the cahow had little time to replenish its stock. As the human population of Bermuda expanded, the colonists, even as they noticed the birds becoming less prevalent, continued to eat eggs and roast birds. The decline of the Bermuda Petrel came rapidly, but not without notice. In 1616, a decree was passed to stop "the spoyle and havock of the Cahowes," perhaps the first-ever conservation legislation in the world.[2] But Bermuda, still a wild and woolly place of jungle islands, couldn't enforce the legislation. By the year 1620, only eleven years after continuous human habitation began, most written accounts made no more mention of the bird. After thirty-seven million years of "harsh howling" in the wind, the bird, which had come and gone from this tiny volcanic archipelago since the Oligocene epoch, disappeared.

In Strachey's time, no one knew what kind of bird the cahow was; early colonists left no drawings or studies. When Bermuda began to be used as a military base in the 1800s, officers and naturalists investigating Bermuda's natural history—without four-color tomes or a clear taxonomy—surmised that the cahows of early writings were the Dusky Shearwater, a similarly shaped seabird found off Bermuda's shores. Others believed the cahow of yore was the Black-capped Petrel, the Audubon's Shearwater, or perhaps a species of auk. Nocturnal and pelagic—it lived mostly at sea—the bird's habits, breeding cycle, lifespan, and even exactly what it looked like, remained unknown. Most agreed if it was a unique species, it was now gone. A Yale biologist summed up the majority consensus when he declared that the bird had gone extinct as early as 1625.

Yet over the years, mystery birds kept turning up, specifically around Castle Harbor in Bermuda's east end. Throughout the 1800s and 1900s, fishermen working the east end of Bermuda reportedly heard calls in the night during a strong squall, or got half a glimpse in the November dusk. On the north shore they called it the "Pemblyco"; on the south shore the cowhow. Some called it the Christmas Bird, for it was heard at the end of the year. In 1906, Dr. Louis Mowbray, who would become the first director of the Bermuda Aquarium, found a live bird in a hole on one of the Castle Harbor Islands; he classified it as a Peale's Petrel from New Zealand, blown off course. A decade elapsed before an ornithologist realized that Mowbray's live bird was actually the real thing: a Bermuda Petrel. The following year, Mowbray discovered bird bones in the Crystal Caves, a tourist attraction; one skeleton he correctly identified as shearwater bones, but he suspected the other skeletal remains might be the fabled cahow. Fifteen years later, an avian paleontologist confirmed his find, yet no one followed up, because as a documentary recalls, "the muddy trenches of Europe were swallowing up too much of the intellectual genius of the time."[3]

Even after a fledgling hit the St. David's lighthouse in a storm in 1934—and was confirmed by Dr. Murphy, a petrel expert at the American Museum of Natural History, to be the cahow—William Beebe, the man who initiated the famous bathysphere experiments off Bermuda and who sent him the specimen, never bothered to look for a live bird, perhaps because in some of his writings he called "cahows" shearwaters. Petrels hail from the same "tubenose family"—birds that drink water and excrete salt through little tubes in the tops of their noses—related to shearwaters and albatrosses. Beebe wasn't that far off.

As a young boy growing up on Bermuda in the late 1930's, David Wingate knew the cahow's storied history, as well as its supposed extinction. A burgeoning naturalist, ridiculed by his peers for his love of nature, and discouraged by teachers for his interest in birds, he often showed his drawings to Louis Mowbray II, the son of the same Mowbray who found the living bird in the nesthole decades earlier, and who had succeeded his father as head of the Bermuda Aquarium, Museum, and Zoo. Mowbray encouraged the young Wingate's nature studies, and always made him feel as if he was "on the brink of something really important."[4]

Wingate, whose alienation led him further into Bermuda's natural world, wandered into a cave one day, and found a set of skull and wing bones. He recalls his epiphany: "I climbed up out of the cave . . . I looked up . . . in the distance I could see the Castle Harbor islands shining on the horizon. And the hair just went up on the back of my neck . . . it seemed impossible that the cahow had been able to survive, but I had to believe it if I wanted to find it alive."[5]

Motivated by his belief, he built a kayak in the summer of 1950 with the help of some friends, then paddled across the open ocean until he finally reached one of the islets in Castle Harbor. Bermuda's cantankerous seas prevented him from landing his fragile craft on the sharp limestone. He vowed to return the following spring. What he didn't know is that he would return, but alongside two of the world's most acclaimed avian naturalists.

Although the outside world had forgotten about the cahow, one man hadn't: Dr. Robert Murphy, the petrel expert and renowned ornithologist at the American Museum of Natural History in New York. Murphy and Louis Mowbray II (the man who encouraged Wingate's studies and who also believed the bird might still exist), risked their reputations to do one last thorough search. Ornithologists look for rare birds all the time. But this was different: how could a cahow population be plausible if it needed forest cover and dirt to burrow in? Bermuda's building boom had begun: roads crisscrossed hitherto wild places, and houses sprouted all over the archipelago. If the bird still existed, it was likely offshore somewhere in Castle Harbor. But during World War II, in an agreement with the American military, some of the Castle Harbor islands were filled in to join the mainland, to form an airport. If cahows nested there, their burrows would have been paved over. The remaining islets in Castle Harbor consisted of hard limestone no bird could dig into. So how could a self-sustaining population of birds still be alive? In scientific circles, the idea of a search provoked skepticism and even ridicule.

Murphy and Mowbray still believed the bird just might have survived in the rocks somewhere. After he secured grant money, Murphy flew to Bermuda in January 1951. Mowbray suggested they bring the fifteen-year-old Wingate along, and Murphy agreed; a surefooted boy could more easily scramble around the limestone. They set a date to begin their search — Thursday, January 25—when the birds might be in burrows incubating. But Wingate, stuck at Saltus Grammar School, couldn't participate until the weekend. In a stroke of magnificent fortune for Wingate, a January gale blew through; the two men were unable to launch in Castle Harbor. On Sunday, January 28, 1951, the seas calmed, and this time, with young Wingate on board, they set out.

Science is competitive, and the bird, if it existed, so rare, they used code words to disguise locations, calling the first island they approached "Island A." Finding numerous crevices with seabird excreta in the openings, they chose a burrow with fresh footprints in front of it. They dug it out partially until they spotted a bird in the darkness: could it be the fated cahow? It took them hours to fashion a noose on the end of a bamboo pole in order to reach the bird; finally, Mowbray fished the pole in, snared the bird's leg and, with a gentle tug, pulled it out into the light. The bird flapped its wings and blinked in the sunlight. Murphy scooped it up and held it aloft. It only took a few seconds of recognition. "By Gad, the cahow!" he exclaimed under his breath—a phrase repeated in newspaper headlines around the world.[6] The bird settled and bowed its head, as if paying homage to its future saviors. Photographs show its vanilla forehead, its ash-colored back, its curved beak, its wings stretched open.

Their discovery was akin to finding a live pterodactyl. The bird did more than emerge from the limestone, it emerged from nearly three hundred and fifty years of extinction. They eventually found eight breeding pairs in the Castle Harbor islands. Endemic to Bermuda, the cahow, they would learn, came there only to breed, only at night, and only to four remote islets.

It was a watershed moment in Murphy's career, and Mowbray's too. For Wingate, as he stood watching Murphy hold the bird up to the sun, the course of his life became clear. There on the edge of the limestone cliff, as the sun beat down and the turquoise water glistened all around, Wingate decided to devote his life to saving the cahow. Some say the seabird saved Wingate's life as well.

In response to the discovery, Bermuda's government set aside the Castle Harbor islands as a nature reserve, but it was almost too late. The cahow was rediscovered just as it teetered on the brink of true extinction. The birds faced dire conditions: forced off of the mainland by humans and unable to dig holes in rock, they lived in whatever holes they could find, often competing for nest sites with the more aggressive Bermuda Longtail, a black and white bird resembling an angel crossed with a flying rat. The longtails returned to Bermuda in March about the time cahow eggs hatched, and the larger birds often pecked petrel fledglings to death in order to claim the shallow cliff-faced burrows. The cahow has one chick per year, with precarious survival rates. At one point less than ten fledglings existed in the world.

Wingate, who helped the recovery project as he completed high school, and took over after college, believed the cahow population might die off for good, until he discovered ten more nesting pairs that—by supreme stroke of luck—had found long, deep crevices in the rock that longtails avoided. But the cahow faced other challenges: rats swam over to the islands and ate their eggs, and when fledglings took their inaugural flights, with nothing to climb up on, they jumped off of cliffs, often bouncing off rocks, landing in the water.

Since the cahows lost their habitat, Wingate decided he would recreate it for them. On the outlying rocky islets, he constructed cement burrows, what he called "government housing," for the cahows. The cement igloos had long, curved entry tunnels to keep out light. Each had a lid to make observation of the nesting chamber easier. He installed custom-made baffles, pieces of wood with a small hole in the middle, on all burrow entrances to keep out the longtails. He then constructed smaller igloos, the size of upside-down serving bowls for the longtails, to entice them to nest elsewhere. Wingate's focus soon turned to Nonsuch, the largest, most protected Castle Harbor island, a place he believed might make the perfect cahow colony.

On maps, Nonsuch Island appears shaped like a question mark, and indeed one might wonder why anyone would want to go there. Craggy, windswept, with steep rock walls covered with moss and small trees, the landing consists of a literal shipwreck: a long wooden boat lies submerged as a makeshift dock. The naturalist William Beebe sank the boat in the 1930s as a tank for his fish studies.

By the 1950s, the fifteen-acre island resembled a blistered mess. Grazed on by cattle and goats, tunneled by rats, covered with invasive plants, the island resembled an ecological junkyard. The only forest cover was provided by the pewter skeletons of cedar trees killed by an imported scale insect, which decimated forests across the archipelago. Nothing about the island brought to mind its name, which meant "unparalleled," a name given by an early governor because there was "Nonsuch like it in the entire realm."

Nestled among the invasive vegetation were buildings left over from Nonsuch's use, first as a yellow-fever quarantine station in 1914, and later as a training school for delinquents: a hospital, a slab-walled mortuary, a dormitory. Wingate, having witnessed the building boom on the mainland while growing up, saw the potential for Nonsuch, not just as a future cahow nesting site, but also as a living museum, "encompassing all of Bermuda's threatened flora and fauna."[7] It was there that he set out to create a miniature, pre-Columbian Bermuda.

After receiving permission from the government, Wingate ripped up the Brazil pepper and fern asparagus and replanted over 800 cedars in the hope that some of them would become resistant to the insect infestation. He imported casuarina and tamarisk trees as temporary windbreaks. With the help of volunteers and a bulldozer Wingate created multiple habitats: rocky coast, coastal hillside, upland forest, freshwater marsh, beach dune, and salt water marsh/mangrove habitat. He drowned invasive cane toads in buckets, poisoned rats, planted prickly pear, buckeye, buttonwood, and yellow-wood; he reintroduced the West Indian top shell, killifish, and the White-eyed Vireo, Bermuda's "chick of the village."

He and his wife, Anita, moved into one of the buildings on Nonsuch, setting up house there so Wingate could be closer to the shy, nocturnal, sea-dwelling birds. Then he camped out on the cahow breeding islets night after night, in the hopes of recording when the adults came in to switch positions with their partner on the egg, yet saw nothing. He knew from his frequent visits that the birds were incubating, changing with their partners every ten or eleven days. But even up close, the bird remained a mystery. That's when Wingate nicknamed it the "ghost bird,"—even to him, it remained as spectral and cryptic as a ghost.

One day, as he prepared to leave one of the breeding islands and return to Nonsuch, a hard wind came up. Wingate signaled the Coast Guard that he would stay there until the weather subsided. He ended up stranded without food or water for several nights. That evening as the sky grew black and a storm raged, he finally saw the petrels descend en masse to switch places with their partners. He had, by virtue of being stuck on the island, unlocked one of the cahow's central mysteries, revealing why early explorers most often heard the "diablos" in bad weather. From their feeding grounds up to four hundred miles away in the Gulf Stream—where upwellings of shrimp and squid awaited—the cahows used the wind energy of the storms to get to their nest burrows and back again. By returning to their burrows most often during storms and on dark moonless nights, the seabirds remained virtually undetected.

Wingate's years of ceaseless data recording eventually revealed the petrel's exact timing for breeding, egg-laying, incubation, and fledging. He knew when the adults returned each November and when the last chick departed in June. Sometimes, because the fledglings had nothing but low scrub to climb, in order to take off they would climb up Wingate instead, taking their inaugural flights out to sea from the top of his head. Petrels mate for life, and he came to know each breeding pair intimately.

Over the years, every challenge solved brought new ones: DDT and other chlorinated hydrocarbon compounds caused unhatched eggs to crack prematurely in the 1960s; light pollution from nearby NASA and US Navy installations prevented them from breeding; one off-course and hungry Snowy Owl gorged on fledglings until Wingate, after much agonizing, shot it. Climate change brought rising seas and increased storms, which flooded some of the lower-lying burrows. Summer started arriving earlier, by a week on average, and late-season chicks sometimes suffered heatstroke.

But under his tutelage and with the help of many volunteers and scientists, the birds prospered. Wingate published papers, spoke at conferences, gave Nonsuch tours to school groups, and campaigned in Washington to end the use of DDT. He negotiated to get airport lights turned off during breeding season. Wingate's young family—by then he had two girls—considered the cahows family. From only eight breeding pairs when first rediscovered, there were eighteen nesting pairs by 1962. By 1985, those numbers more than doubled, and by 2000 fifty-nine pairs existed.

The cahow still faced many problems. First, biologists feared handling the rare birds, afraid the creatures might die of stress. And after Nonsuch was restored to lush pre-Columbian splendor, the cahows didn't flock there to nest. They returned to the same rock-strewn islets year after year, islets which had little usable habitat, islets which would eventually be destroyed by hurricanes, to lay their eggs. Meanwhile, Nonsuch Island, a perfect bride for the cahows—with forest cover, soil, and protection from hurricanes—sat ignored, waiting in vain for her bachelors.

Getting to Nonsuch requires taking a fishing vessel from the Bermuda Biological Station for Research, steering out into the sea around the curve of Bermuda's tail, then veering into Castle Harbor. My first time there, spray drenched the bow as we chugged between enormous angled rock formations. Flying fish, little silver torpedoes with bat-wings, skipped across the water. A squadron of Bermuda Longtails appeared, trailing their ropelike tails behind them.

Wingate—a tall, broad-shouldered man with a head of gray curls and a potbelly—met our small group, a few tourists and myself, at the dock. He toured us around as if he was just the caretaker, not the founder, of one of the most remarkable ornithological recovery projects of the twentieth century.

Following Wingate through the forest was like being thrown back in time. Dense forests of Bermuda palmetto, southern hackberry, Bermuda olivewood, and cedar grow next to a shimmering freshwater marsh seeded with purslane, hyssop, and smartweed. The forest here, created with a muted palette of endemic species, grows nowhere else in the world. It was this Bermuda that shipwrecked sailors saw in the early 1500s, suffering from the stink and scurvy of their voyage, wading the final several hundred feet to shore. The smell of hot wood, decaying palmettos, sagebrush and dogwood, mingled with their perspiration and sour breath.

We followed Wingate out of the woods to a clearing. Breakers hit the rocks below; the sky remained gray and the wind full: a storm was coming. A normally shy man, he grew animated as he showed us, using his arms as surrogate wings, how the birds use their hooked beaks and toenails to scramble to the tops of small trees, a habit formed during their evolution when there was no visible ground to take off from. When they reach the top of the branches, they fling themselves into the wind.

Although we couldn't see them, the birds hovered in the air before us, creatures of the wind, of the night. Standing amidst the lush foliage of Nonsuch, it was easy to imagine the entire island filled with cahow burrows, to see their oneiric silhouettes streaming overhead at midnight, slinking into the ground to feed their chicks.

At one point that day, I looked at Wingate as he stood poised next to the graceful silver skeleton of a cedar tree. The task of helping the cahow survive seemed tenuous. At times, he worked himself to exhaustion for the sake of a handful of birds. He sacrificed time spent with his children when they were young, missing out on much of their childhood. The goal of increasing the breeding pairs to one thousand over the next several decades seemed a Sisyphean task. Even, looking at it from the outside world, a ridiculous one. What did it mean to keep a small cadre of rare seabirds alive? Why did the cahow matter?

Two years later I returned to Bermuda in the hopes of finally seeing a ghost bird. I was moving from Boston to Seattle, and didn't know when, if ever, I'd be back. At first glance, the island hadn't changed much, even after Hurricane Fabian in September 2003, one of the most powerful storms recorded there. But if you knew where to look on the mainland, you saw boulders lifted by the sea like chess pieces, resting on top of cliffs. In some places, overwash cleaved the faces off the escarpments, leaving them blank and unknowing, a tabula rasa for the next storm. Fabian felled about half of the large trees on the mainland, mostly non-natives. Nonsuch, with new endemic forest cover that could withstand Bermuda's climate, suffered only a few upturned palmettos.

In the Castle Harbor islands, Fabian wiped out 60 percent of the cahow nesting sites; storm surge had submerged the islets. The hurricane, like a crazed octopus, hurled the burrow lids—manhole covers designed to fit over the nesting chambers for easy access, weighing nearly fifty pounds each—onto the sea bottom. It broke burrows in half, filled others with debris, and cleaved whole islands. Thankfully, the petrels had yet to return for breeding, otherwise they would've all been killed. Wingate and Jeremy Madeiros, his former second-in-command who had taken over as Conservation Officer after Wingate retired, tried to rebuild some nests with concrete and cement. But many of the sites were gone, unable to be rebuilt because the cliffs had sloughed off into the sea, as if the island had shrugged off a cape.

On Cooper's Island, now joined to the mainland, cement picnic tables lay buried in sand and the spine of an old military road heaved and cracked, the work of some supernatural chiropractor. Like everywhere in Bermuda, international detritus lined the beach: a child-sized plastic car, a fire exit sign, planks, oil barrels. I spent a few hours there one evening, trying in vain to hear the bird, though they rarely call after a certain time in the season.

I found its keepers as elusive as the bird itself. Wingate was off-island. I tried to contact Jeremy Madeiros, to ask if I might go out with him during his monitoring forays, but he hadn't returned my messages or e-mails. Even people who knew Madeiros had difficulty reaching him. My cell phone didn't work on island, and I had only a pay phone at my room. I wasn't sure if my timing was great, or horrible, but it was all I had, a few days at the end of March.

I was scheduled to go out to Nonsuch again the next day. The dock there, more treacherous after the hurricane, proved too dangerous to use, so weekly public tours stopped as a result, but I arranged to get there with a college group staying at the Biological Station. We would anchor off South Beach, don wetsuits, and swim in. It was suggested I might run into Madeiros there. Although he was nowhere in sight, seeing Nonsuch again on a sunny day in all its profuse splendor felt heartening.

The next day was my last in Bermuda. Rather than go to Crystal Caves where the cahow bones were found, I hopped a pink bus to the Bermuda Aquarium, Museum, and Zoo. I had an appointment with the curator of the natural history collection. Before I went upstairs to meet her, I visited the small building that housed endemic wildlife. I found what I was looking for near a tank of Bermuda rock lizards: recordings of local birds. I took the small phone off the hook, and, before I pressed the button for "cahow," I forced myself to listen to the Yellow-crowned Night Heron and the White-eyed Vireo, even though I knew what they sounded like. I wanted to savor the moment. I had read a description of the cahow's sound in a children's book: "*aaw-eeh aa-aaw-eeh, keeh-eek-eek-eeek.*"

I pressed the button for "cahow." I heard the sound of waves and then, suddenly, a long, low, drawn-out, addled whoop, ending in a sharp whistle upwards. Sailors and settlers called it "weird," "strange," and "demonic." To me, the calls sounded like a serrated moaning, rising in pitch at the end to a whistle, a sound that twisted between mourning and passion, building then ebbing like the waves around it.

Staff at the museum knew I was trying without success to get hold of Jeremy, but they managed to get his wife on the phone. She told me he was out on the harbor, working, but he always checked in with her. She also knew I was leaving the next day.

"Perhaps he can pick you up and take you out this afternoon." She promised to call me back, but I didn't want to get my hopes up.

Upstairs, I sat with the curator in a wide, white back room, filled with light and stuffed bird specimens. The boulder-sized skull of a short-finned pilot whale, from a nearby dive site, sat on the floor. A gigantic Short-tailed Albatross, about three times the size of a Canada Goose, watched over us from a high perch. His pink, curved beak, similar in shape to the cahow's and from the same tubenose family, resembled a small banana.

On the table sat several miniscule containers of cahow bones; they looked ethereal, as if from a real ghost. She selected a wooden box that held a delicate curved skull. The specimen was marked "found as a juvenile on June 16, 1963," the secret location where it was found written in David Wingate's neat handwriting. She placed it into my palm: the skull, just under three inches and the color of weak tea, possessed the curved bony protrusion where the keratin of the beak fit onto it. There had once been a million of these birds. What remained for scientific study seemed paltry by comparison: a few boxes of bones here and there in museums, some dead specimens.

We pored over several more boxes, then I followed her into a windowless chilled room lined with open metal trays that pulled out of the wall, like trays at a bakery. Instead of pastries, the trays held dozens of "skins"—birds preserved for study, their soft organs removed. She pulled open a tray at eye level.

There, stuffed with cotton or excelsior, lay a half-dozen dead cahows, adults and fluffy chicks, with white gaps instead of eyes. They resembled limp missiles. One was stretched out like an angel, on its back with wings spread as if crucified. It reminded me that the word "petrel" is derived from St. Peter, who supposedly walked on water, from the bird's habit of "dancing" just above the waves.

Their bodies, streamlined for flight, though similar in size to pigeons, were nothing like pigeons. Sculpted by wind, they were yielding yet rigid missiles, their wings powerful arcs of feathers. Evolution does not preclude beauty. Their white foreheads faded into gray and their backs were washed with darkness. Their tails were gray skirts brushed with milk, and their black beaks ended in a graceful sharp hook. With eye-slits close to their beaks, their faces appeared childlike. The crucified petrel had one of the cahow's singular field marks: a gray smudgy "thumbprint" visible on the outer part of the underwing. Another revealed pinkish legs, ending in dark webbed feet. A dark gray fluff ball of a chick lay on its side.

I could have stayed there longer, but the call I had been waiting for came in: I was to meet Madeiros at the Tuckerstown dock at 4:30 p.m. A few minutes later, a taxi left me at the end of a narrow road, on a small picturesque bay. Soon a battered motorboat came around the bend, bearing the logo of the Department of Conservation: a large cahow tilted into the wind with wings outstretched. At the helm stood forty-seven-year-old

Madeiros. I rose to my feet as the boat glided dockside; he greeted me as he took off his green, wide-brimmed canvas hat and ran a hand through his black hair. His manner, like Wingate's, was humble, accommodating, and polite, like a waiter at a tea house. He began piling some empty gas cans onto the dock. "Wingate was so focused, we'd be out there and run out of gas, so now I always bring extra."

I couldn't believe my luck: after helping me aboard, he guided the miniscule boat out into the harbor, and explained he was taking me with him to monitor chicks on some of the more difficult islands. The boat, about five feet by fifteen, might easily capsize in a large swell, but was ideal for maneuvering up to rocks. The sea glowed green as we rode over it, the sky's dome a dusky blue overhead. A Bermudian accent lilted through his words as he told me how he came to be a keeper of the ghost bird.

Madeiros, who studied plants in college, became an apprentice in the government's horticultural division. They farmed him out to Nonsuch, where he worked with Wingate, and began assisting with monitoring the cahows. "I learned that everything was holistic; the birds needed the islands as much as the plants did." After eighteen years, he eventually became Parks Superintendent; after Wingate's retirement in 2000, Madeiros became the second-ever Conservation Officer in charge of the cahows. Initially he felt the great weight of keeping the cahow alive. "Then I was so busy I didn't have time to worry," he said. Because the cahow program continued its success, his anxiety gradually lifted. There were seventy-five breeding pairs now, up from fifty-nine when I last visited and when he took over. Roughly two hundred birds existed today. He was monitoring the entire breeding population of a species.

Madeiros squinted, one hand on the wheel, the other running over his moustache. I asked him what it was like when he first touched the ghost bird. "It was like holding a dinosaur, like holding a legend alive and in the flesh." Madeiros likens the cahow not to the dodo or the Great Auk, but to the legendary phoenix that rose from the ashes. "I felt then, and still feel now, the privilege of being able to work with a species that's such a keen survivor."

In the open harbor he pointed out a distant fortress, dating from 1612: the Castle of Castle Harbor. Once out in the channel, Madeiros increased our speed. As the boat skimmed the water, he noted that they often had weekly gales between December and April, some hurricane force. Usually waves came over his bow. "If I can get out here two days a week, I'm lucky. I usually have to rush out, do my checks within six to eight hours, and then get back to shore." Madeiros hadn't been able to get out to the islands since the previous Friday—my timing couldn't have been better. We would be performing "chick checks," weighing them and taking other data and checking the bird's health.

We approached an atoll and Madeiros nosed the boat toward a ledge. To the inexperienced eye, the island offered nowhere to land, but Madeiros threw the thirty-pound anchor overboard, grabbed his backpack and, when the boat crested the swell of a wave, jumped onto the ledge, tied the boat off onto another rock, and helped me ashore. Wingate often swam to the islands in rough seas to monitor chicks, and now I understood why. The island wasn't more than thirty feet above sea level, and consisted of pocked limestone no bigger than half a football field. He pointed out Fabian's damage: boulders lay strewn about, lifted like bottle caps by the waves that had washed right over the island. It was hard to imagine anything lived here, yet if you looked, life abounded: below the water line, West Indian top shells clung to the limestone, and a low cover of scrub survived on the island's top. As I followed him to higher ground, objects resembling rocks materialized into the removable nesting burrow covers. We squatted down next to an entrance hole, covered with a baffle to keep out Tropic Birds. In the sand in front of the entrance, Madeiros pointed to several webbed footprints, barely visible. The chick had a feeding visit sometime during the last few nights. The parent had come, regurgitated, and left. He took a small notebook and pen out of his backpack, removed the cement lid—four inches thick and the diameter of a basketball—and beckoned me to look inside.

Peering into the darkness, I made out a grapefruit-sized gray wad of fluff. He reached down and brought it up into the sea air. The chick protested, making "weep weep" noises, but Madeiros supported its feet and preened its cheek, and it settled into his hand. Covered with two layers of dark gray down, the only thing that showed out of the fluff were its liquid black eyes, and its famously curved and hooked beak. "I call them dust bunnies." He settled it into a cloth bag and weighed it.

I've worked with bird banders before, and spent many hours watching before I was allowed to handle and release even the most common birds. Madeiros lacked any proprietary airs. He held the chick out to me and I cupped my hands together. "Why don't you hold the hope of the next generation?" I supported the chick's underside, and it nestled into my hands. The wind ruffled the down on its body. Every few seconds the chick, which was between three and four weeks old, flapped its wings, little V-shaped appendages covered with fuzz. I was holding one of thirty-seven chicks in the entire world. As I held it up to eye-level, its eyes darted around my face, then receded into a wild place.

As the wind whipped at our clothing, Madeiros spoke about the real weight on his shoulders, his plan to move the birds to Nonsuch. He spent months in Australia working with Nick Carlile, an expert on the endangered Gould's Petrel, a bird that climbs ninety-foot trees in the rain forest in order to take flight. Carlile successfully moved the Gould's Petrel and started another petrel colony nearby, tripling the endangered birds' population. Carlile taught Madeiros that handling petrel chicks would not stress them. Madeiros spent a month in Australia banding petrels, albatrosses, and penguins, and learning to feed petrel chicks that have been moved (once they were moved, the parents couldn't find them to feed them). He started the banding program in Bermuda as soon as he returned.

Madeiros explained the central mystery of why the birds hadn't voluntarily nested on Nonsuch. He nodded at the chick cupped in my hands. Petrels imprint on their birthplace. Once they imprint on the rocky islands of Castle Harbor, they return there to nest. To get the cahows to return to Nonsuch, Madeiros would have to get the chicks imprinted on Nonsuch. Adults that hadn't imprinted as fledglings wouldn't lay eggs there.

But just how the chicks imprint turns any notion of human superiority upside-down. When the chicks grow to a certain age, the parents desert them. Instinct propels the young birds outside at night, and they prepare for life on their own by exercising their wings in the darkness. Simultaneously they look up at the star-filled sky, and see, for the first time, the heavens at the exact location where they were born. This behavior is captured for the first time ever, in the documentary film *Rare Bird*, and is spooky to watch: the birds fixate on the horizon and bob their heads, scanning from side to side. This memorization of the sky is the equivalent of a global positioning system; after spending their first four to five years of life entirely at sea, *Pterodroma cahow* comes back to its birthplace, often landing within a few feet of the burrow where it hatched.

In order to get the birds to nest on Nonsuch instead of the rocky islets, Madeiros moves the fledglings during the day, in covered tomato boxes. For the past several years, Madeiros has translocated thirty-five of the healthiest chicks, hoping they will imprint on the stars from hurricane-proof Nonsuch. Volunteers help him feed the birds during their final weeks because their parents can't find them, and then do night watches to record when the birds take off. In 2004, the first fledgling cahow came out of a burrow, looked intently at the stars on the horizon, and took off from the forested cover of Nonsuch, the first time in centuries that a cahow had fledged there.

Before we placed the bird back in the hole onto its bed of straw and sand, Madeiros showed me its webbed feet, and the tiny nails on the end of each toe. "They still haven't lost their digging impulse, even though for the last couple of hundred years they've been living in rocks instead of soil. Sometimes the little ones just scrabble around in the sand." I said goodbye to the dust bunny, and Madeiros placed it into the hole. We continued on to the other burrows. Life came forth from the rocks: fleabane ruffled in the wind and cockroaches scuttled. Bermuda rock lizards roamed inside the burrows, eating insects. We took data and scrambled from nest to nest, recording the color of guano left by the adult birds, which indicated their diet: squid, crabs, or fish. Each fledgling cheeped as he lifted it out of the nest, and we felt the warm swollen stomachs full of regurgitated sea-life. Soon we were done. We picked our way down to the boat, and headed to the next island.

The island we landed on had been cleaved into three separate parts by Hurricane Fabian. I followed Madeiros, and jumped onto a ledge of slippery rock, then scrambled up a limestone bluff using my hands. At the top, the land looked as if a giant machete had split it. He pointed down into a gully, which used to be solid ground, and the home of numerous burrows.

"We couldn't even rebuild those," he said. "There was nothing to rebuild on."

The fifty-foot-high waves of Fabian had washed over this island as though it were a pebble: a whole section vanished. If the hurricane had occurred during nesting season, the fifty-five-year cahow project would have drowned. As it is, all the nesting islets will eventually be destroyed by the increasingly violent storms that hit the archipelago.

"You're standing on the famous 'Island A' of the rediscovery," he told me.

Murphy and Mowbray had both been here; I imagined them holding the bird aloft in the famous photograph, and the thought of it raised hairs on the back of my neck. We took more data and weighed chicks, until there was one nest left. Pointing to another gray limestone bluff separated from us by water, Madeiros told me an adult bird was incubating a failed egg over there. If I wanted to see an adult cahow—a bird so rare that only a handful of people have seen it and handled it, a bird most native Bermudians have never seen, even though it's their national bird—this was my chance.

We slid down a steep embankment to sea level, keeping the rocks at our back. A tilted boulder made a makeshift bridge over to the rest of the island, but the sea washed over it at intervals. We waited, then jumped. We scrambled to the base of the cliff and climbed up. The sun slanted from the west, and the limestone shone white. The wind buffeted us. I was shivering. The rocks plunged straight down to the ocean.

Madeiros found the nest entrance under a ledge. He pointed to the sand in front of the hole. The webbed footprints pointed in, but none pointed out. "There's an adult bird in there, alright." I kneeled next to him as he lifted the concrete lid off. Then he plunged in his arm, and pulled out a full-grown adult cahow by the back end.

There are times in your life that become touchstones, that define all else that came before and all else that will follow. Nothing prepared me for this living dinosaur: the vanilla forehead, fading into charcoal, the liquid black eyes, the torpedo-sleek body, sturdy but soft. The dark feet, set far back on its body, ending in the webbed and nailed toes that evolution gave it. It glowed in his arms. I was face to face with the ghost bird, one of the rarest seabirds on the planet.

He flipped the bird onto its back, and cradled it. "This one's unbanded," he said. I reached out to stroke its stomach. The sun, the cliffs, and the wind all seemed part of the bird. The bird bent its neck up and bit my hand. We laughed.

"That's your souvenir," he said. Then he slipped the bird into a soft bag and had me hold it while he retrieved a band. I cradled it upside down against my chest. The cliff fell away to the sea not six inches from my right foot. I was holding a pterodactyl in my arms—I was holding the dodo, the auk, and a thousand other birds, all the birds that had evolved and gone extinct before Bermuda was first discovered. I lowered my head and inhaled the musky scent of its ancient body, a soft, sour, not unpleasant odor.

In his essay "The Lives of a Cell," Lewis Thomas says "[I]t is illusion to think that there is anything fragile about the life of the earth; surely this is the toughest membrane imaginable in the universe, opaque to probability, impermeable to death. We are the delicate part, transient and vulnerable as cilia."[8] Madeiros told me that they were tough survivors, but I hadn't grasped it until now. I sensed its tenacious heart, its resolve to survive. I understood the cahow was as tough as the earth—it would go on, despite us. And that we—with our cars and wars and money and petty concerns and selfish genes—remained wasted and trivial.

Madeiros applied a band to the bird's leg, measured the wing, and then held the bird against his chest, facing me, and spread each paddle-shaped wing to show me the characteristic thumb print near the tip of its primaries on the underside, just like the famous rediscovery photograph. I understood why Murphy said "By Gad, the cahow." As I looked at the bird's face, this legend turned toward the setting sun, I felt the weight of belief. I understood why the cahow needs to be saved. It has nothing to do with celestial secrets, with its beauty, with its beating heart. The bird isn't just inextricably linked to us by mystery and awe. It is us. Like Thoreau's Wood Thrush, it's a "part of our unfallen selves."

I grew up reading *National Geographic*, *Popular Science*, and the Time Life books about deserts, evolution, and rain forests. But it was a cloth-covered guide to North American birds that attracted me when I was eight years old. I carried it around much of the time. I can conjure up the worn blue cloth cover in my hands, and the smell of the binding's glue. I studied it, flipped the glossy pages almost daily, burning the shapes and names of the species into my subconscious without knowing it: the Indigo Bunting, the Great Blue Heron, the Rose-breasted Grosbeak. The birds took on the significance of talismans; when I looked at them, I felt longing and attachment, possibility and despair. They represented an escape into some unknown future, a place "out there" that, once reached, would fulfill me somehow.

In birds, unlike humans, even the drabbest of creatures has categories—drawn and laid out in the pages of the book—and import.

The dull brown sparrow comes in so many variations: Rufous-crowned, American Tree, Clay-colored, Chipping, Vesper, Field. Each had a paragraph, and a tiny map. The guide was like a yearbook where even the ugly, unpopular creatures were given focus and loving detail: the flamboyant *Aix sponsa* and *Cyanocitta cristata* next to *Passer domesticus*. They had a grace and beauty that I felt I would never achieve; in them lay the hope of my possible transformation from invisible preadolescent to, if not something beautiful, to something like the Roseate Spoonbill: a vision of awkward loveliness. I studied those pages hard, drawn by their allure, always emerging somewhere else. To paraphrase Robinson Jeffers: did it matter whether I hated myself? I loved the wild swan.

We headed back to the mainland. Sunset painted the sky violet and apricot. We passed the ancient fort, and Nonsuch. Longtails cavorted above us. Bermuda's archipelago, silhouetted in the sun, looked primordial and resplendent. I told Madeiros he had the best job in the world. He nodded and grinned.

"I adhere to the rule of intelligent tinkering," he said, "which is to say: never throw away any of the parts." He continued, "We like to think that we are lords of creation, masters of our fate, and we're not. We're one of thousands of species sharing the planet."

I asked him how he would feel if the cahows were suddenly wiped out. As he piloted the boat over the waves, his face belied a determined optimism, the same optimism I had seen in David Wingate.

"Failure is not an option. They're a tough species. Our goal is a thousand breeding pairs, which may not happen in my lifetime, but it will happen. If we can do it with this, we can do it with anything."

As we rounded the limestone lip of Tuckerstown harbor, Madeiros added, "I was depressed about Bermuda's development. To be able to address it . . . that's one of the highest callings. We're a very selfish species . . . man is incredibly capable of destroying. What I found is that a few people in key places can make all the difference."

Bermuda's a microcosm, an isolated archipelago, a suburb in the sea, a miniature Earth. Plastic garbage swirls in the Sargasso from the world over; it erodes into tiny balls that sea turtles ingest. They die, their intestines impacted with plastic.

A staffer at the bio station told me, "We're hell bent on destruction. The more affluent the society, the less people give a damn. All the lovely Bermuda cottages are disappearing, and in their place huge houses erected. There's a habit of everything being destroyed . . . with the cahow, it gives you hope." A common refrain on the island: if the currency of nature is spent, we will be poorer. Whatever happens to the cahows will happen to us.

Nonsuch remains an island of extremes: the dense and the sparse, the noise and the calm, destruction and rebirth. The keepers of the ghost bird continue their earthly tasks. Wingate shoots the Kiskadee Flycatchers that eat the Bermuda rock lizards, even though he has officially retired. Madeiros, on Nonsuch with his family, will continue his cahow night watches. His six-year-old daughter will help him take data as he feeds this year's crop of translocated chicks before they take off for sea.

In a few weeks, a new set of chicks on Nonsuch will emerge and face the stars for the first time. They'll stare at the horizon as if hypnotized, bob their heads, scan and somehow fold the positions of the galaxies into their ancient brains. If they survive their first five years at sea, they will return to Nonsuch, and in doing so, will join a group of the first cahows to live there in 400 years.

Chapter 18

The Hour (or Two) before the Dawn

Donald Kroodsma

"3:00 a.m. 6-28"—so declares the lighted watch face when I punch my alarm off. With sunrise nearly three hours away, I slip out of bed, and I'm soon out the motel door and on my fully loaded touring bicycle, heading west out of Wisdom, Montana, eager for yet another small bite out of our listening journey from the Atlantic to the Pacific. At whatever speed I choose this morning, stopping where I please to look and listen, I will play my way across the sagebrush from here to the forest eleven miles away, and then ascend to Chief Joseph Pass, twenty-five miles to the west. My son and riding companion will follow later, and we'll meet up at the Pass.

"Best not to get up before the sun"—I smile to think of David's simple declaration as we listened during that first dawn chorus back in Jamestown Beach, Virginia. Since then, he's written almost daily in his journal something to the effect that "Dad's had yet another giddy bird outing

---

Numbers preceded by ♫ refer to recordings at ListeningToAContinentSing.com.

before sunrise," whether out biking or prowling around the campground. David has come to tackle the early morning on his own terms.

He's twenty-four, beginning a new life on this cross-country ride, having just finished his master's in earth systems science at Stanford one term early so that he could begin biking with me in early May, celebrating Virginia and Kentucky during the birdiest month. He says he's riding with me for the free bike, the new touring bike I bought him just for the purpose, as he's eyeing an epic ride from the Bay Area in California down to the tip of South America in a few years.[1] A scholar, musician, athlete—he does it all, just last spring captaining Stanford's ultimate team to a national championship. Best of all, he's my son, and he's carried me on this trip, with his flair in the "kitchen," his upbeat "It'll all work out" through bicycle repairs and my three emergency room visits (leg issues!), to this our 56th day on the road together. I increasingly feel I'm half the man he's becoming, and I couldn't imagine a finer riding companion.

I'm almost fifty-seven, searching for a new beginning myself, trying to "freshen up" from an academic position that's gone south. As a professor, I've taught mostly ornithology; as a scientist, I've researched birdsong with a passion, studying how songbirds learn to sing, how they use their songs and large song repertoires, song dialects, and so much more. I reasoned that a cross-country ride would enable me to hear and celebrate all I've come to learn. Now, almost two months into full immersion, everywhere I hear birds saying "Come sing with us," and I am increasingly realizing with each day that I will abandon my guaranteed-for-life professorship at the University in favor of a life listening with birds. Already I feel *FREE*![2]

With no moon, the constellations burn intensely against the still-black sky, the Milky Way ablaze across the heavens. Off to the right lies the North Star, and I trace the Big Dipper, the Little Dipper, the Dragon. My eyes then sweep across the heavens to spot old friends who fly there, especially the Swan. Mars glows red off to the east; Venus and Jupiter are below the horizon.

As I cross the Big Hole River just out of town, a lone Marsh Wren (♫ 294) sings lazily off to the right. A *western* Marsh Wren, with all of the harsh buzzes and rattles and whistles one would expect from his more than one hundred different songs, so different from the eastern Marsh Wren we heard on the other side of the Great Plains. What's in his head to make him sing now in the dead of night? Is he a young bachelor who sings for his first mate, or perhaps an older male who already has one or two females in his "harem" and is greedy for more? I yearn for a second singer to engage him, the two of them matching each other song for song in one of the most impressive singing duels to be heard across the continent, but none rises to the occasion.

From beside the road a surprised Killdeer flushes, *kill-deer kill-deer kill-deer* (♫ 136). High overhead, snipe winnow in earnest (♫ 276), perhaps a dozen of them within earshot, unseen meteors hurtling earthward with the wind gusts from each wingbeat whistling through their stiffened outer tail feathers, *wuwuwuwuwuwuwuwu*. Some call intensely from the ground, too, *tick-tock, tick-tock, tick-tock tick tick tick*. Have they been active like this all night? When do they rest?

To the north, beacons of shimmering, greenish light dance from the horizon up beyond the North Star, engulfing the Great Bear. They shift to the left, then to the right, now a half crown, then a full ghostly halo over the earth's pole, pulsating as if alive. The northern lights, the aurora borealis, what a gift now in late June.

The constellation of lights that is Wisdom recedes behind me, and I cruise along effortlessly at a modest six miles per hour, listening . . . to my breathing, as I'm climbing a gentle grade . . . to the whispered concert of my pumping legs, the drive train, the two tires rolling on pavement . . . to the sound of darkness swishing by my ears . . . to the very idea of sun's first light and the dawn chorus racing toward me from the east, at a clip of about eight hundred miles an hour . . . to my voice whispered into the night, "What a spectacular time of day to ride!"

In the vast stillness of the sage, with not a whisper of wind, from all directions I soon hear the sound of sagebrush awaking, and a little after 4:00 I stop to marvel. Seemingly thousands of tinkling bells announce Horned Larks in dawn song (♫ 284); nearby, I listen as one male hurries through four to five notes each second, every half minute or so punctuating his effort with a rising, stuttering flourish, only to begin all over again. At a blistering pace, he races among his more than two hundred different notes and perhaps ten different flourishes. Later this morning, he'll offer only occasional flourishes with long pauses between them, a different bird altogether after the dawn rush has passed.

Sage Thrashers (♫ 217). Some have been singing at a rather leisurely pace all night long, I know, but just listen to them now! Each has hundreds of half-second song snippets that he strings into songs a minute or more long. From the nearest singer, I listen for the mimicry . . . for now, he tells mainly of the Western Meadowlarks and Brewer's Sparrows who will eventually awake here, but it would be hours of singing before he tells all he knows.

Coyotes yip and howl in chorus off to the right. Far away, a lone cow bellows, but nearby something snorts just across the roadside fence. A bit spooked, with the weak light attached to my helmet insufficient to reveal the source, I quickly move on, past a Mountain Bluebird singing aloft to my left (♫ 260). High overhead, he warbles softly, twenty gentle notes over ten seconds, a five second pause, and then he's at it again. The larks, the thrashers, now this bluebird—in darkness, with most songbirds still quiet, each greets the coming day in its own special way.

With just enough light to see, I pull into the overlook at the Big Hole National Battlefield. In the serenity of this dawn, I strain to imagine the mayhem here on another dawn back in 1877. Chief Joseph and roughly 750 Nez Perz Indians are camped down below, where the skeletons of tepees now stand in memoriam along the river. Horses graze nearby in the lush meadows. In a surprise attack, the army soldiers first disperse the horses and then advance on the teepees, firing low into them. Fierce fighting erupts, and the soldiers retreat to the grove of trees that I see just above the river, where sniper fire from the Indians pins them down. As the snipers contain the soldiers into the night, the Nez Perz bury their dead and depart, continuing their flight from where they began in Oregon to where they'll eventually surrender just short of Canada almost two months later.

Sobered, I bike on over the river, listening to songs that the Nez Perz themselves probably heard on that fateful dawn. Sandhill Cranes sound off, he bugling low, she finishing the duet high, the pair sounding as one (♫ 272). Yellow Warblers race among a dozen or so different songs, filling all air time between songs with frenetic chipping (♫ 309), such a different bird than the one who will emerge after sunrise, when just one special song will be repeated at a far slower pace, without the chipping (♫ 94). How different the yellowthroats, another warbler, each male with just one version of his *wich-i-ty wich-i-ty* song that he'll use throughout the dawn and day (♫ 295), and now the Northern Waterthrushes, again each male with only one song that must suffice for all occasions (♫ 287). I pause to listen: *sweet sweet sweet swee wee wee chew chew chew chew,* yes, rapid and emphatic, three-parted, descending from one phrase to the next, just like all Northern Waterthrushes everywhere. Warblers, I reflect, each singing in its own way, with such variety among them.

"Thank you, lady bird!" I give credit where it's due. Although the female of most songbird species doesn't sing, she chooses who will father her offspring, and by choosing a male who sings to her satisfaction, the females have over evolutionary time orchestrated all of the singing patterns that I now hear. And perhaps it is during the dawn chorus that the females are listening especially intently, so that males pull out all stops then, singing with all the energy and talent they can muster so as to make a good impression. Hurrah for the aesthetic tastes of the choosy females: "You are the silent composers of this symphony."

Just beyond this North Fork of the Big Hole River lies the forest and the beginning of my gradual ascent to the Pass. A little after 5:00 a.m. now, about forty minutes before sunrise, the birds here are in full song, the dawn chorus at its peak.

How empty a dawn would be without the energy and enthusiasm of flycatchers. They had backseats when the music was being passed out, once said noted ornithologist Frank Chapman.[3] Since then, we've learned that the brains of flycatchers and songbirds differ in striking ways. Songbirds have special clusters of neurons that guide song learning; young birds babble as they learn to sing much as we humans learn to speak, resulting in local dialects, large song repertoires, and often highly complex songs. In contrast, flycatchers have no such brain centers for song learning; their songs are innate, with no local dialects and with relatively small repertoires and simple songs. But just listen to them now!

I cycle past pewee after pewee, all westerners, each rapidly alternating his two songs, *tswee-tee-teet . . . bzeeyeer . . . tswee-tee-teet . . . bzeeyeer . . .*, a song every second and a half, forty per minute (♫ 275). These Western Wood-Pewees picked up mid-continent where the Eastern Wood-Pewees (♫ 144) left off, each species with one special song that is held in reserve to herald the dawn. How fascinating, then, the *quick! THREE BEERS!* from the closely related Olive-sided Flycatcher here (♫ 348); he sings at twice the pace now than he will later in the morning, but he has no special song to greet the day. Three species in the genus *Contopus*, all with a common ancestor, each having come to sing in its own way.

Two empids, two "Mosquito Princes" in the flycatcher genus *Empidonax*, sing here as well. I follow along with a Willow Flycatcher (♫ 225), knowing that like Willow Flycatchers everywhere he will have three recognizable songs, because the songs are in their genes, and the song genes are (mostly) the same everywhere: *FITZ-bew* with a *FITZ* of sharp, pure-toned notes; *FIZZ-bew* with a fizzy introductory note; and a brief stuttering, rising *creet*. Which song will be next, and when? I always wonder, as the performance feels disjointed and unpredictable. And the Dusky Flycatchers (♫ 245)—each also has three different songs, though so different from those of the willow. The willow uses his three songs whenever he sings throughout the day, but the dusky seems to concentrate this trio of songs at dawn. Yes, each in its own way.

Overhead I hear the repeated *peent* calls of a nighthawk (♫ 170), punctuated with his occasional booming *VROOM*. I count a dozen *peent*s, then detect enough of a pause that I know what is coming next: He's in a dive, soon in the moment of truth as he extends his wings out in front of him, the wing feathers vibrating loudly, *VROOM*ing much like the sound of a bellowing bull or a very short freight-train. Other nighthawks *VROOM* in the distance, the low frequency of this mechanical song carrying much farther than their higher *peent*s.

Three sparrows: Each White-crowned Sparrow (♫ 269) sings the only song he knows, two half-second plaintive whistled notes giving way to husky, complex notes, slurred whistles and buzzes, all conforming to the song pattern of other males around him; I listen as each weighs in, each saying in effect "I sing the local dialect, therefore I am." Each Lincoln's Sparrow (♫ 290) bubbles away, so energetically, each with his own unique, House Wren–like song that shifts abruptly from one pitch to the next; it seems that he has only one song that he repeats over and over, until for whatever reason he alters the sequence of trills, or adds or drops a phrase, creating a whole new effect on the listener. I pause at the next Song Sparrow (♫ 327), listening to him methodically repeat his song over and over, smiling when he abruptly switches to another of his ten or so different songs. Among these sparrows are no special dawn songs, but I do hear a heightened energy, a faster pace of singing, certainly a more hurried switch from one song to another among the Song Sparrows.

I cycle past a gathering of Western Tanagers (♫ 311). Three birds now sing within earshot, as if they have come together at dawn to contest one another at their shared territory boundaries. How fragmented their dawn effort seems, with the burry phrases given at half the pace of their daytime songs; for the near bird I count, seven burry phrases then a pause for his characteristic call, a dry, quarter-second, rising *pit-er-ick*, then eleven phrases and a call, then five and call. It's his standard dawn performance, indistinguishable from that of the Scarlet Tanager (♬ 38) except for the call, which back East is a unique *chip-burr*.

A doe snorts loudly and repeatedly at me, her fawn just off the road up ahead. I ease on by.

About 5:15 now, still half an hour to sunrise, directly down the road behind me I see a sliver of a moon rising, Venus its companion just behind. I continue west, past Warbling Vireos, MacGillivray's Warblers, water-thrushes, juncos, and a whistled song so sweet but so foreign. I imitate it, trying to burn it into memory, guessing Pine Grosbeak but unsure. A startled group of elk hustle across the creek and into the woods. Slowly I meander up the road, "the most beautiful place I have ever been," I say to myself.

Cycling down the road, I occasionally disrupt a gathering of two or three Chipping Sparrows (♬ 145) who sputter their songs at each other on the pavement; it's all part of their dawn routine, each male leaving his own territory to seek out other males in the neighborhood, to argue intensely, it seems, and beak-to-beak, perhaps with females taking notes on how well they do. And I marvel at how different these western Warbling Vireos (♬ 325) sound from the eastern birds (♬ 124). There's only a hint of the spirited, undulating singsong of the eastern's *If I SEES you, I will SEIZE you, and I'll SQUEEZE you till you SQUIRT!*; these western songs are more choppy, with successive notes rising and falling more rapidly, and without the high, emphasized ending. Their genes differ, too, I have read, revealing that eastern and western Warbling Vireos have been isolated from each other for some time on opposite sides of the Great Plains.

I stop to listen to a robin (♬ 285), walking beneath him. He plays out his mood in a series of low carols, from five to fifteen at a time, followed by one to three high, screeching *hisselly* notes. I pick one unique carol, hearing how he offers it several times among other carols over fifteen seconds, and then it disappears, only to reappear a minute or so later. I linger, to count carols and *hisselly* notes, to listen for recurring carols, trying to detect his pattern . . . Very nice—the essence of robin, so expressive, with so many options as to how to say what is on his mind.

A Veery (♫ 287), another thrush, just beyond the robin. Exquisite, enchanting, a breezy downward spiral of several fluty notes, a bit nasal, as if he's singing into a metal pipe, it is said: *da-veeyur, veeyur, veeer, veer*. Three subtly different songs I hear, delivered in a highly regular sequence, one, two, three, one, two three, the same three songs that he'll sing whenever the spirit moves him throughout the day.

I bike on, but soon brake to still another stop. "A *Great Gray Owl*," I whisper into my voice recorder. "Just ahead on a small post across the road. He's HUGE. Look at that head, so round and enormous; two large facial discs . . . small, beady yellow eyes at their centers . . . Just below his yellow bill is a black chin, with snow-white moustaches extending out to the side, looking much like a black-and-white bowtie . . . feathered legs, impressive talons emerging to grip the post . . . What an extraordinary creature!" As I try to capture the moment, he glares directly at me, no doubt trying to fathom in his own way the odd creature he sees: a two-wheeled beast all in red, running with a two-legged beast all in yellow. Perhaps it is only five minutes later, maybe an eternity, when he lifts into the air ever so silently, a phantom disappearing into the forest.

Announcing sunrise, as hole-roosting woodpeckers seem to do everywhere, Red-naped Sapsuckers now drum and call all about me (♫ 291). Nice rhythm, I think, a few rapid strikes followed by thumps in slower, irregular cadence, so different from the methodical drums of other woodpeckers. The rhythm of each master percussionist here feels much the same from bird to bird, but the resonance varies, depending on the quality of the chosen substrate. Music, I choose to hear.

I climb more steeply now, on a three to four percent grade, and with about two miles to the Pass, the higher elevation is announced by Mountain Chickadees (♫ 267) and Ruby-crowned Kinglets (♫ 265). With less than a mile to go, in a lush fir and pine forest, Hermit and Swainson's Thrushes abound, some of the finest songsters on the planet. The hermit (♫ 334) leaps from one pitch to the next as he delivers his dozen or so songs, with successive songs chosen to be especially different, the enhanced contrast striking; each Swainson's Thrush (♫ 243) spirals his three or four songs heavenward, much as I feel in climbing the Pass.

And a Varied Thrush (♬ 304)! I answer him, whistling and humming at the same time to generate two voices that oddly beat against each other. He jumps in pitch and tone from song to song, and I do the same; together we bound among five different songs before coming back to the first. I'm a lousy Varied Thrush, and he ignores me, but for a brief moment I *am* a Varied Thrush, and it is ever so fine.

I move on, but yet another thrush soon freezes me beside the road—a Townsend's Solitaire (♬ 292). From the treetops out of sight to the right, his songs ripple out over the mountain, clear warbled notes delivered in a great rush for nearly half a minute, and then he rests, but only briefly. Those who tell of this song gush in superlatives, most glorious, most beautiful, the richest and fullest and clearest and sweetest and sparklingest, the finest mountain music.

"Thrushes," I say—what marvelous singers. I tick them off for today, a Mountain Bluebird, the robins and Veeries, and it just keeps getting better, with Hermit and Swainson's Thrushes, a Varied Thrush, and now the solitaire. I imagine each of them in a dawn chorus, all together, each greeting the dawn here in its own extraordinary fashion. What a special lineage of exquisite singers! Just what was it in the blood of the ancestral thrush that has now yielded such magnificence?

Just short of the sign announcing Chief Joseph Pass I am stopped by a Pacific Wren (♬ 300). "Mr. *western* Winter Wren," I hail him, "just listen to you go!" I last heard his eastern cousins atop Mt. Rogers in Virginia, but this western bird is an even more remarkable singer.

And here comes David up the hill! "Buenos días, señor," I greet him, though the words would be an understatement in any language.

"You've been taking your time," he says, smiling, beaming even, standing athletically in the saddle to climb the steeper grade here. What a beautiful image: my son, chief (well, only) cook on this expedition, on his black touring bike (old guys like me need a flashy red bike) with red saddle bags and twenty-seven gears to choose from, tent strapped on the rear, about seventy-five pounds of bike and gear, everything needed to thrive on a seventy-day adventure crossing the country.

I laugh, and we compare our versions of the twenty-five miles we've just biked. "Left about seven o'clock, letting the sun ease me out of bed," he smiles, "a perfect morning. The battlefield was a sad place. Great forest." I tell him some of what I've heard, especially in the last mile.

"Hear this Pacific Wren? . . . the high, tinkling sound?"

"Yeah. What's that all about?"

"Oh, the most spectacular song in North America. It goes on and on and on, maybe ten seconds, and he's very creative in taking segments of his song and rearranging them, making all new songs." I describe how the songs are so high and fast that our human ears can't process the sounds, but slow him down on a computer and then we can hear what he no doubt hears as he leaps about among his tiny, tinkling notes. "Neat bird. And he's only yea big," I gesture with thumb and forefinger.

"Cool . . . but how can all birds be the most spectacular?"

"It's just the way it is," I explain fully.

Finally, Chief Joseph Pass, at 7,241 feet yet another crossing of the continental divide. It's a little after 9:00; over the last six hours I've averaged about four miles per hour, a brisk walking pace.

A most perfect morning. Launching westward in darkness, I drifted by mysterious singers active through the night. The larks were first to announce dawn's early light and the awaking songbirds overtaking me. Soon I was immersed in the full chorus, riding with it, relishing each singer, hearing what was on his mind, the larks, thrashers, and bluebirds in the sagebrush, the flycatchers, sparrows, tanagers, thrushes, and so many more in the forest. Well before the actual sunrise, the intensity waned, the full chorus having advanced beyond me, continuing its sweeping arc around the globe, me happily knowing that it will return again tomorrow. Buoyed all along the way by unseen songsters, I glided effortlessly on my flying machine through time and space, today, through the early morning hours over the twenty-five miles from Wisdom to the Pass, but over eons, through evolutionary time and space, slipping from one lineage of birds to the next, recognizing each species and each lineage as a snapshot in earth time. I marked their differences, their similarities, exhilarated at how all this came to be as I mentally traced lineages back to their ancestors, farther back to the origins of birds, back even farther those countless millions of years to the ancestor that eventually gave rise to both humans and birds, when we were one.

Overhead, a flock of Red Crossbills (♬ 252) calls as they bounce along in the sunshine among perfect spires of subalpine fir. Yes, a perfect morning.

Chapter 19

Secret of Owls

K. Bannerman

Above our heads, the evening sky is the same delicate blush pink as a rose petal, but the air bites into exposed skin with icy fangs, and the conifer branches appear black against the snow. There is nothing gentle, sensual, or romantic about this hike. Tonight, the wilds will be cold and unforgiving. We know this, going in, but in an academic way.

An hour along the trail, we come to know it in a visceral way, too. We know it in the same way a deer knows to avoid a hungry cougar. We feel it instinctually in the marrow of our bones, this imminent death, this frantic discomfort. Our hearts beat a little faster as our fingers tingle in their wool mittens.

This winter has been uncharacteristically crisp. Normally, it's marked by endless rainy days, gray skies, and frost in the early mornings, but this year, the temperature has plunged to record depths, and the deep snow that would normally be restricted to the ski hills has spread all the way to the ocean shore. Our snowshoes help us shuffle along the forest path, avoiding drifts and circumnavigating stumps, spreading our weight and buoying us up; this would be an impossible trek without them. Thick marshmallow gobs of powdery snow cover the tree boughs and reflect the light of the rising moon. Everything is bathed in a whimsical blue. Where the day's sun melted the snow into a crust, one can catch a silvery glint that skitters over the surface like a school of sardines. A little crystal icicle garnishes the tip of each cedar branch. When a breeze ruffles the high canopy, a veil of white flakes shimmers and floats effortlessly to the ground. These details are delightful, but we don't bother to point them out. In fact, we speak very little, because it takes concentration to move forward, and we're listening intently for our quarry.

Cumberland Forest stretches up from swamps and wetlands, along ridges of hemlocks and pines, into a series of canyons that contain frozen rivers and a couple of small, ice-locked lakes. We're intimately familiar with this landscape, Ella and I—we have lived at the edge of these woods for almost twenty years, and spend most summer afternoons in the shade of its second-growth trees. But tonight, under the light of a wolf moon, slogging through drifts of snow, I almost don't recognize these pathways. In the summer they are welcoming and lively, but tonight the paths are lonely, brooding, introspective. They hide riddles. They plot against us.

And why shouldn't they? We're outsiders. Look at us, wrapped in wool scarves and goose-down jackets with squeaky synthetic shells, as plump as a couple of astronauts bounding over the lunar surface! We're so well bundled against the elements, the only way to tell us apart is by the difference in our height. Ella and I have become vaguely-human-shaped bipeds shuffling down the incline from the trail, cutting across the solid waters of the upper wetlands, weaving between clumps of dead cattails and golden reeds. By the time we reach our destination, our blood has quickened with the exercise. Our breath spills into the night air like white feathers. We're both smiling.

Allen Lake is a round dollop of water between a series of low hills. At the north shore is a bench, hewn from a single log with a chain saw. It offers a wide view southward across the lake, and I swipe the snow from the seat with a single swing of my arm as Ella puts down her pack, removes an old blanket roll, and unfurls it along the bench to protect our bottoms from the moisture. As we sit, she pulls out a flask and two tin cups.

"Here," she says quietly, not wishing to disturb the peace. She hands me one.

I hold it still as she pours out the tea, then pours a cup for herself, and we sit side by side, sipping Earl Grey.

How have we come to find ourselves here? Why are we not curled up with books by the fireplace, sipping our tea in relative comfort?

If you nurture a fondness for nature, maybe you'll understand: for the last three nights, we've heard the sounds of Vancouver Island Northern Pygmy-Owls. From the edge of the village, gazing towards the mountains, I've tried to pinpoint their location, but it's been difficult. The *hoo-hoo-hoo* echoes around the ridges and canyons. It bounces from the hillsides, a soft call that plays tricks on one's ears. Are they up by the lake, or down in the wetlands? How many are there, hiding in the trees? It's not uncommon to hear Barred Owls or Great Horned Owls in Cumberland Forest, but the unique call of a pygmy-owl is relatively rare, and lately, they've been more active than usual, as if they're trying to tell me something urgent.

That's foolish, of course. They have no secrets in particular that they wish to share with me.

But their chorus has sparked our curiosity. Northern Pygmy-Owls were once numerous on the island, but since the 1920s, their numbers have been declining fast due to logging, urban development, and other activities leading to habitat destruction and so the fragmentation of their population. In fact, I've never seen one in the wild, though I've heard them from time to time (and certainly of late, I've heard them every night!). This protected forest is one of the last places where they can live in relative security. Over the last twenty years, the people of a nearby village have banded together to protect these lands, and Cumberland Forest is one of the few untouched places where these tiny gray-brown birds could possibly thrive. Here there are plenty of old trees that have been scarred with holes by woodpeckers, holes the pygmy-owls need to build nests. Here there's a fine buffet of small rodents and songbirds to munch upon. Here an endangered animal could live a life unimpeded by human progress—this forest could be their little owlish kingdom, if they wish.

Logic dictates that more hoots mean more owls. I hope, fingers crossed, that a viable population has discovered these woods. I hope to glimpse an owl, and set my heart at ease that maybe, just maybe, they've come to stay.

The bench is not a comfortable place to sit, but the blanket helps. We keep our eyes on the trees, our ears open for any hint of a hoot. If we hope to see a pygmy-owl, we'll need to be sharp: these tiny puffs are no larger than a House Sparrow, with a pair of blazing yellow eyes that pierce your soul. I sip my tea as my gaze ranges over the dark hump of the far shore. We'd be hopeless without the moon, but lucky for us, it's full and bright, with nary a cloud to hide behind.

"Almost as bright as day," says Ella. "Isn't this lovely?"

A movement to the west. Ella lowers her cup. I hold my breath.

A black-tailed doe takes delicate steps as she strides from the wings onto the center stage of the frozen lake. We watch, breathless, as another emerges, then another. We watch them and remain mute until all that's left are parallel tracks in the snow, disappearing into the trees on the opposite shore.

"Isn't that lovely?" Ella whispers.

"Mmm," I reply over the curve of my cup.

Minutes stretch out. The cold begins to seep in between the cracks in my armor—the exposed space between mitt and sleeve, the back of my neck where my scarf gaps. I feel Ella's body pressed against my own, but neither one of us breaks our gaze to start a conversation; this human connection of touch is simply a reminder that we are not alone in the darkness. Our eyes scan the upper canopy, searching, as we lean against each other. From the perspective of a beast watching us from the woods, we have cleaved into a single statue, perched on the bench, doing nothing more exciting that staring into the distance.

Then.

"Oh!"

Ella's surprise escapes as a puff of white mist.

"What?" I try to keep my voice low, but she surprised me, and I speak a little louder than intended.

"Didn't you hear it?"

We both fall silent. I strain to catch whatever she heard, and soon, I hear it, too.

Hoo-hoo-hoo

Starlight glints off our wide smiles.

And then:

hoo-hoo-hoo

hoo-hoo-hoo

hoo-hoo-hoo

All around us, a choir of tiny popcorn voices, filling the air. We smother laughter behind our mitts.

Ella leans to my ear and whispers, "What are they so excited about?"

"Maybe mating season?" I suggest, although all my research tells me it's too early for those sorts of shenanigans. And maybe it sounds crazy, but I detect a note of crisis in those little voices. One mustn't anthropomorphize, I know, but I hear it: a warning.

"Don't they sound . . . sort of . . ." I choose my word carefully because Ella is pragmatic: "Aggressive?"

She cocks her ear, as if this little motion will help her decipher the meaning. "Um . . . maybe?"

Hoo-hoo-hoo

One chirps out right behind us in the bush. Both of us leap in our seats. Ella squeaks.

And instantly, the chorus stops.

She slaps her mitts over her mouth and says in a muffle, "Did I do that?"

The woods fall inert. Every motion ceases. From the gentle breeze in the reeds to the whisper of branches, it all stops. Never in all my life have I felt such a weight in the air, as if the silence is water, and we are floating under the surface, pressed down by the salt and the sea. Suddenly Ella and I are two tiny figures in an immense black-and-white photo. No, this isn't Ella's doing—her squeak couldn't have this much power. The whole world has fixed itself in place like a wide-eyed rabbit paralyzed by the sudden flare of a flashlight. My heart starts to thump in my chest, and I want to run but I can't. All I can manage in those few seconds is a quick glance at Ella's face, pallid in the moonlight, as a flush of fear rushes over her features. She feels it too: this is wrong. There is something coming in our direction. A malevolent force prowls amongst these trees, and the hoots we hear are the warning, given by pygmy-owl to pygmy-owl, as the unnamed evil approaches.

Stay still, stay silent, pray it passes you by.

And then, like a wave, it comes.

Soundless, deadly, dreamlike, it cuts through the darkness towards us as it skims over the lake's surface. It is a mighty ghost in the blackness, a huge smudge of milky white in the gloom. And it's fast, so fast! We have only a moment to glimpse it, to try and catalogue these alien features, try to slot it into our expectations. Over the reeds and the ice it soars directly towards us, then at the last moment, the phantom swoops up and over our bench, so close that I can feel the gust of air created by the downstroke of its powerful wings.

For an eternity, we are rooted in stunned surprise.

Then Ella barks, "Holy moly!" and that makes me snort out an awkward laugh. The tension snaps. We both heave a sigh. She fumbles for words, and finally stutters, "Was that . . . was that . . .?"

"I think we just saw a Snowy Owl!"

We sit for a moment in wordless adoration.

"Wow," she replies, because really, what else is there to say?

Both of us stand. We know nothing else the night gives us can compare to such a sublime vision. Ella rolls up the blanket and tucks it back in the sack along with the cups, then heaves the bag over her shoulder. Our snowshoes hush on the crust as we trace our steps along the path that heads homeward. With this unseasonably cold winter, marked by bitter temperatures and polar conditions, I suppose we shouldn't really be surprised that a majestic hunter of the Arctic might find itself so far south, drawn by plentiful game, exploring new territories, extending its range while conditions allow. The pygmy-owls are (with a little luck) here to stay, but this transient visitor has given us an experience that we'll never have again, and for that, my heart is grateful. I feel like we've witnessed a mythical creature, an icy phoenix with ebony talons and feathers made of frost, and its presence has left me breathless.

On the opposite shore of the lake, safe in their little woodpecker holes, the midnight chorus starts again, and this time, it sounds joyful and chipper, sweet and happy. For tonight at least, the pygmy-owls can relax. Danger has passed.

Hoo-hoo-hoo

Chapter 20

Guardian of the Garden

Renata Golden

The red coachwhip wound around the black locust tree, its head inches from the entrance to the bluebird box. At almost five feet long, the snake had climbed fully off the ground, its neck arched to poke into the hole. Inside peeped two nestlings.

I had been watching the birdhouse outside my bedroom window for months. Western Bluebirds overwinter in Santa Fe, and some cold winter mornings I watched four or five males, females, and juveniles pile out of the box like clowns out of a Volkswagen after spending the night huddled together for warmth. In April the antics of the eager male, quivering his wings and gamboling on the branches of the locust tree to the amusement of the judgmental female, woke me most mornings. By early May the new couple shot into the entrance hole of the birdhouse with remarkable precision, carrying building materials of twigs, grasses, and feathers. While the eggs lay in the nest, warm and cozy underneath mom, dad brought a caddis fly or lacewing home for dinner. I first heard the tiny voices around

Memorial Day. By then both adults hunted nearby, carrying insects to feed the hatchlings with increasing intensity.

I enjoyed watching this little scene of domesticity, perhaps because it was so entirely missing from my life. I fall into the census category of "childless," a lackluster term that defines a woman not by what she is but by what she doesn't have. The definition of the term skirts the issue of whether a woman finds herself in this state as a result of choice, biology, economics, philosophy, or simple procrastination. Except nothing about reproduction and its aftermath is ever simple.

Bluebirds are quiet birds, but the bright June morning the snake appeared, both parents complained loudly enough to catch my attention even as I sat at my computer on the other side of the house. The birds flitted around the snake but weren't big and mean enough to actually attack it. And yet they made their intention clear.

Let's take a minute to contrast the beauty of the bluebird and the menace of the coachwhip. An ancient battle fraught with symbolism was being fought in my high desert garden. In Navajo culture a bluebird is an allegory of creativity and creation, but a snake communicates with the supernatural. In Christian religions the snake is blamed for the mess we've inherited, all because of a woman's desire for more. I maintain a healthy respect for snakes and find them fascinating, beautiful, and maligned. On the other hand, I am an active advocate for bluebirds—I clean out the nest box each fall, put out dried mealworms in spring, and toss pillows at menacing Scrub Jays during nesting season. But I was being tested in my own little Eden. Contestants on an ancient battlefield were about to collide, and my sympathies were being called out. Even inaction counts as action, especially regarding life and death in a nest box.

I grabbed a garden hose and blasted the snake. Startled, it slid up the branches of the locust tree until both it and I realized it could either return down the path that it had taken or fall on my head. The coachwhip was smart enough to choose the first option and spiraled back to the nest box. I continued my offensive, nervous that the snake would try again to get inside the box, if only to get away from me. I also worried about how to spray the box without getting its contents wet. I had a split second to balance defending the babies against possibly soaking them. When the snake wound its way around the box again, I inadvertently soaked the box.

As I pressed my thumb harder over the end of the hose, I tried to justify my choice. I was only doing what every other privileged human has been doing on the planet for millennia—playing God, deciding who deserves to live and who should go hungry, my choices propped up by rational arguments about declining species numbers and subjective aesthetics. But my logic was skewed. What did I know of the conservation status of red coachwhips in the desert Southwest? (IUCN: Adult population size unknown, current population trend stable.) Was I taking charge of my environment out of habit, out of control, or because it felt good? Was I just another example of everything out of whack in the world?

The answer depends on who is looking. Consider the constant noise about the sanctity of motherhood, its undisputed joys, and the argument in this country over whether women should even be given that choice. When I was growing up on the South Side of Chicago, all the women I knew were either mothers or nuns. Neither role appealed to me, precisely because they seemed to be my only two choices. Except for the two weeks in grade school when I tried to hedge my bets about my "calling" and attended after-school vocation classes led by Sister John Kathleen, joining the convent was as unpleasant to me as childbirth. If an aunt or cousin handed me a baby to hold, I would develop a sudden need to do something—anything—else. I belong to the 40 percent of American women who have never given birth. I never tried to feed an ailing infant or keep an altricial newborn safe. I am proof that motherhood does not come naturally to all women. In fact, when birth control failed, twice, I terminated not one but two pregnancies—decisions not easily made at the time but about which I have no regrets. I am not alone in this sentiment.

The coachwhip hit the ground and raced toward the back gate, where I assumed it had slipped out of the courtyard. But fifteen minutes later the snake's red head peered like a periscope over the rock steps leading down to the gate. A second male bluebird, part of an extended family group from what I assumed was the previous year's brood, joined me and the parents, who were still perched in the tree, noisily defending the nest. Together, aided by a garden hose, we made the snake feel unwelcome, our vigil lasting more than an hour. After repeatedly sticking its neck above the rocks only to be shot in the head with a hot stream of water, the snake finally disappeared.

Determined to be sure we were rid of the threat, I called a neighbor, who showed up with a snake stick fashioned from a long-handled hedge trimmer, packing foam, and duct tape. We disassembled the rock steps that led to the gate but found no trace of the snake. I spread a band of rock salt around its pathway for good measure. I have no idea where I got the idea that rock salt could be an effective barrier of any kind—it isn't. It served only as a measure of my desperation and helplessness. It also underscored where my loyalties lay, although in the heat of the battle I never once questioned my aversion to the snake. Perhaps because, unlike my bluebird box watch, I had never witnessed the snake family's struggles to survive and had never met its cute little snakelets. Or maybe because snakes have never in popular white American culture been associated with happiness, but rather the opposite. Coachwhips have an undeserved reputation for aggression, when they actually dislike humans as much as we dislike them, and can race away at speeds up to eight miles an hour. Despite their unfortunate name, they will not whip you to death with their tails, although they will jump at your face if cornered. They find rodents, lizards, and other snakes as tasty as birds, and I knew the snake could easily find a snack in the green space on the other side of my courtyard wall. As my neighbor and I reassembled the rock steps to the back gate, I trusted the snake to remain gone. I felt satisfied as the guardian of my garden.

Around lunchtime a warm breeze picked up, which I hoped the coachwhip would find annoying. I peeked at the box through a window, surprised to see one of the nestlings sticking his head as far as he could outside the entrance hole. He was straining so hard his tiny claws were gripping the edge of the hole. Within seconds he had gotten his body wedged into the opening until he suddenly popped out and onto the horizontal branch that stretched in front of the box. The wind had grown stronger, and the nestling clutched at the branch, tottering front and back like a drunken sailor on high seas. He soon lost the fight against the force that would one day sustain him, as a wind he couldn't negotiate blew him off the branch.

I now had one fledgling on the ground, one nestling still in the box, and a hungry snake on the prowl.

The fledgling displayed amazingly astute survival skills—staying almost invisible from the neighborhood hawks and owls; emerging from hiding places behind lilac bushes and tall grasses just long enough to beg for food; confidently hopping onto rock walls when one of his parents arrived with lunch. He grew stronger. No snake appeared. One morning after he had spent three days on the ground, I found him standing in the middle of a round patio table. He seemed as perplexed as I was about how he had gotten there and what to do next. Maybe he was contemplating the awesomeness of aerodynamics. Maybe he was astounded by his new freedom. Or perhaps he was preparing for the huge leap he was about to make. We looked at each other for several long seconds before he flew, full of grace, to the low branches of a nearby peach tree.

Outside the courtyard wall the parents had already begun building a new nest in a second nest box. This box sat atop a wooden post designed to deter a climbing snake. I'm sure they intended this box to remain unmolested by Scrub Jays and coachwhips, but I wondered what else they were planning. Ornithologists say that to the parents the value of a brood is directly proportional to the cost of replacing it. This little family had already experienced significant loss. About a week before the first nestling fledged, I watched the female struggling inside the nest box, pulling and tugging at something obviously heavy. After wresting it free, she pushed it through the entrance hole and carried it off. I knew it was a newly hatched bluebird, but I didn't know if it was alive or dead or dying. She landed in a juniper tree on the other side of the courtyard wall, minus the baby. I traced her flight path, fruitlessly searching the ground and the heavens to understand what had happened. Did the baby die after hatching? What killed it? Maybe it was still alive. It was barely a day out of its egg. Why would a mother carry off a living baby? Were they moving to the other nest box? Had they been disturbed? Was I to blame? The next day I watched the male do the same thing, but this time I found what he dropped. The baby bird, eyes closed and naked except for a few wisps of what looked like hair and the beginnings of pinfeathers on wings smaller than its head, lay under a lilac bush. It showed no signs of injury or attack. Its translucent skin revealed an abdomen swollen with the yolk it had swallowed before hatching. It was crusty, tragic, and dead.

Biologists had been saying that with the drought New Mexico was experiencing in 2013, only 25 to 30 percent of fledglings would survive, and 90 to 95 percent of the birds that did survive after fledging would die before they were old enough to reproduce. This added up to mean that only 22.5 birds out of a hundred would make it past their first year. Of the four eggs laid in my nest box, two babies died before fledging, one had successfully navigated its way to the peach tree, and one remained an unknown in the box. Four days had passed, but the second nestling hadn't left home. The parents still came to the nest box bringing dinner, which Bluebird Baby #2 willingly ate, but it was evidently less eager to follow its sibling out into the world. I wondered if my spray had indeed prompted Bluebird Baby #1 to jump too soon. The strong, confident behavior he displayed outside in the courtyard gave me hope he would survive, although to my knowledge, I never saw him or the snake again.

On the fifth day after the first bluebird fledged, Bluebird #2 popped out of the nest box, miscalculated the horizontal branch, and tumbled to the ground. She immediately displayed a personality different from Bluebird #1. She seemed hesitant, alone, and afraid. She stumbled when she tried to hop up a step. She didn't hide under the lilac bushes but remained out in the open and cried nonstop. Dad with a bug in his beak appeared on the courtyard wall only once, watched the baby cry and beg, and flew off, still gripping the bug. Mom only occasionally glanced baby's way while continuing to build the new nest in the box beyond the wall.

I listened to the baby bluebird cry for two days, unanswered, as it begged for food. The more it cried, the angrier I grew. My anger had its own evolution. At first I couldn't understand all the work the mother had put into this little family only to give up on a life barely begun. I found it hard to accept that she had made a commitment to lay the eggs, incubate them for two weeks, hunt for food, and defend against predators, only to stop short of giving every last one of them all the energy, attention, and resources she could afford. My anger was critical and accusing and directed solely toward the mother, busily building the new nest fifty yards away and showing no interest in the life already here now. I began to doubt the infallibility of the instincts that drive birds and animals through the patterns of their days. Did this mother really know what she must do?

"Hey!" I shouted at her once as she flew to the other nest box. "Somebody's hungry over here!"

But then, what did I know about being a mother? By my reckoning, maternal instincts are not innate. When, as a teenager, I tried to envision my life as an adult, I never saw kids in the picture. I adore puppies and kittens but feel no connection to newborn humans. Although I have always smiled and pretended to admire babies and pregnant women, I figure that the planet is already about to collapse from the burden of carrying us all. I don't need to contribute to that weight.

I was watering the hollyhocks early the third morning when I spotted her. A fuzz ball of spotted down and pinfeathers, Bluebird #2 was standing with her head tucked under her wing, next to a bush less than ten feet from me. I was so startled I spoke to her.

"There you are," I said, surprised that she made no attempt to move away. At seven o'clock in the morning the June sun had been baking the ground for an hour. The fledgling had not taken refuge from the heat or the predators. But at the sound of my voice she picked up her head and looked at me with an expression of absolute sorrow. I moved slightly closer. "Are you okay?" I asked.

She tried to move and tipped over. With barely the energy to spread one wing to break her fall, she opened her beak and began to pant.

She lay not far from the locust tree. Her nest box now sat quiet and empty. She had only been out of her shell for maybe three weeks; she had only been out of the nest box and on the ground for two days. As I saw it, those two days were the most forlorn, cruel, and empty days.

I put down my watering can and sat on the ground in front of the bird. There was nothing I could do but bear witness to her existence, even as it ended. "I'm so sorry," I said. We sat together in the sun for twenty minutes, until the bird collapsed and breathed its last breath, and I went back into the house.

When I tell friends the story of the bluebird, they ask me questions I can't answer. Why didn't you save it? Why didn't you pick it up and put it in a box and bring it to the Wildlife Rehabilitation Center? Why didn't you do something—anything?

The truth is it never entered my mind. I was so busy being indignant at the mother's behavior, I never considered intervening. I can make excuses—I had other things to do, I was working and couldn't pay attention, I wasn't always home. I had no idea how to care for an abandoned fledgling. The Norwegian writer Karl Ove Knausgaard has said that "indifference is one of the seven deadly sins, actually the greatest of them all, because it is the only one that sins against life."[1] Was I indifferent? Was this mother? Were we both sinners?

I needed answers. I called the Cornell Lab of Ornithology and spoke to a woman named Robin. Why did this have to happen? Had a virus infected the nest box? That explanation was plausible in New Mexico and could account for the deaths of the first two birds that hadn't fledged. Perhaps Bluebird #1 was resilient enough with a strong-enough immune system to flourish in spite of the virus, but Bluebird #2 was just less fortunate, gene-wise. Was something wrong with Bluebird #2 that I couldn't see? Maybe I had damaged Bluebird #2 with my blast from the hose. Maybe she never would have made it anyway. Or maybe this scenario was playing out according to rules I would never be privy to.

Robin explained that parents need to balance the demands of their fledglings against the maintenance of themselves and the species. In purely biological terms, the parents make an investment in the future. Increasing one baby's chance of survival—the ultimate goal being, of course, reproductive success—comes at the cost of investment in other offspring. But, I asked Robin, how do the parents do that math? At what point are the parents giving more than they can afford?

The parents analyze their offspring and rely on cues to determine the chances their babies have of being successful, Robin explained. Mathematical models suggest that parents allocate more resources to bigger, stronger, and more colorful babies when resources are scarce. But, I argued, I had enough mealworms in trays to keep the entire family sated for weeks, and there seemed to be no dent in the lacewing population. I craved answers that Robin didn't have. Did the mother stop feeding

Bluebird #2 because she knew it was going to die? Or did it die because she stopped feeding it?

The calculation that a female bluebird must make is simple. Her single-minded goal is the success of future generations; her vision reaches beyond herself. I once mentioned to a friend—another childless woman—that one day I might regret not having children, because who would care for me in my dottering old age? That's a lousy reason to have a kid, she replied.

Terry Tempest Williams has written that to become a mother is an unspoken agreement to be forever vulnerable.[2] A thesaurus tells me that the opposite of vulnerable is impregnable. I know that I am neither. My math is also simple—I am the sum of my decisions, only one of which was to not have children. A selfish decision, maybe, that was actually made out of generosity.

Like the snake in the garden that tempted the first woman to make a decision, my coachwhip taught me that every decision has consequences that live for generations. I am certain that the bluebird mother would agree. When the nest in the box outside the courtyard wall hatched a second brood later that summer, she had not forgotten the snake as much as made new, better decisions—better investments in her future.

In my heart, I keep the memory of Bluebird #2 close.

Chapter 21

Chasing the Ghost of the Imperial Woodpecker

Tim Gallagher

I have a recurrent nightmare. It usually goes something like this: I am traveling alone on a rough footpath through the Sierra Madre of northwest Mexico. Even though I'm dreaming, I know right where I am, because I've studied the area for so long. It's a dangerous place and always has been, where for centuries various rebels, outlaws, and other desperados have gone to lose themselves—or more often to find other people to terrorize. It is a land of drug growers, smugglers, and kidnappers. But none of that enters my mind. I'm on a quest to find an Imperial Woodpecker, and nothing can distract me.

As I trudge ever deeper into the forest, the pines become larger and larger, lofty old-growth giants with trunks several feet across, towering into the sky. Then I hear it: the deafening raps of a foraging woodpecker, and it's far louder than I ever imagined, each blow like an explosion echoing through the forest, again and again and again. It's like nothing I've ever heard. What else could it be but an Imperial Woodpecker—the largest woodpecker that ever lived? And I'm racing as fast as I can after the sound, flying across the ground as fleet as a deer. Suddenly a dark-haired man dressed in black, holding an AK-47, steps out from behind a tree, blocking my path, and I cry out in shock.

On the good nights, I wake with a jolt right then, shivering in a cold sweat. But other times the dream continues, and the man is shouting at me in a tongue I can't decipher. Is it some unusual mountain dialect of Spanish—or perhaps the ancient Nahuatl of the Aztecs? I try speaking to him in broken Spanish, in English, and even in sign language, but he says nothing, glaring at me with burning eyes—pitiless and menacing. Pulling out my wallet, I show him my identification . . . some pictures of my kids . . . some American money. He knocks them to the ground and shoves me hard with the butt of his rifle, driving me away from the trail, deeper into the woods. I walk ahead of him, sick with dread. Finally he stops and points the rifle at me. I glance down and see his finger tightening on the trigger.

And then I hear it . . . a loud noise like a toy trumpet, ringing out high above me. Looking up, I see an Imperial Woodpecker hitching up the trunk of a lofty pine, and the bird is huge, like a big white-billed raven with gleaming white on its lower back. The bird pauses and lifts its head, gazing at me with a yellow eye, its brilliant red crest blowing shaggily in the wind. And it's the most beautiful sight imaginable—so vivid, so real, like I could reach out and touch it—and I choke up. An instant later, a powerful explosion rips my vision, and all is darkness.

I often wake up yelling and thrashing around when I have this dream, frightening my wife and making it difficult for either of us to get back to sleep. I'm not sure what it means. I've faced many dangerous situations over the years in my quest to observe and study birds, climbing lofty cliffs to reach the nest ledges of Prairie Falcons, Peregrine Falcons, and Gyrfalcons; canoeing through swamps and bayous alongside floating cottonmouths to search for Ivory-billed Woodpeckers, but I had never before felt this kind of subconscious dread—and it got worse each time I ventured into the Sierra Madre. A malevolent cloud seemed to hang over everything I did in Mexico, and I wondered, Why am I doing this? It was a question often asked by my wife and children, and I never had a good answer.

Of course, I wanted to know whether this remarkable bird still lives, and, if so, whether it could be saved from extinction. No one else was looking for it anymore. And the bird had virtually never been studied.

A handful of scientific papers—written mostly by people who never saw one—and a few dozen specimens are the only record that this species ever existed. For the Imperial Woodpecker, there was no Jim Tanner—the ornithologist who studied the Ivory-billed Woodpecker intensively in the late 1930s. I feel bad about that, but is finding an Imperial Woodpecker really worth dying for?

Perhaps it's no surprise I became so thoroughly obsessed with the Imperial Woodpecker. I'd already spent years searching for its closest relative, the Ivory-billed Woodpecker, and actually had a sighting canoeing through an Arkansas bayou in 2004. A year or so later, I was photographing woodpecker specimens at Harvard's Museum of Comparative Zoology when an assistant curator brought me a tray containing more than a dozen Imperial Woodpeckers. I was stunned by their beauty and majesty. They easily dwarfed the ivory-bill.

For millennia, the imperial's pounding drumbeat echoed like the blows of a wild axman through the old-growth forests of the Sierra Madre as it bored into the massive, grub-infested pines, hammering on them powerfully for weeks at a time until they groaned, shuddered, and finally toppled with an impact that shook the ground. Victorian ornithologist John Gould dubbed it the Imperial Woodpecker. But it had already been named long before. To the Aztecs, it was *cuauhtotomi*; to the Tarahumaras, *cumecocari*; to the Tepehuans, *uagam*; and to the Mexican Mestizos, the *pitoreal*.

Large as a raven, with the deepest black plumage and brilliant, snow-white flight feathers looking like a white shield on its lower back, the Imperial Woodpecker was impossible to miss. Its eyes glowed golden yellow; its massive, chisel-like bill shone white as polished ivory; its feathery crest curved forward to a shaggy point. And it was noisy, blaring a loud toot like a toy trumpet as it hitched up a pine trunk, foraging for beetle grubs as big as a man's thumb. But its showiness was its undoing—that and the fact that it stayed in tight family groups and often hung around, curious, when one in its group was wounded or killed.

Norwegian explorer Carl Lumholtz, who spent eight years traveling on muleback down the 900-mile length of the Sierra Madre Occidental in the 1890s, declared the birds would soon be exterminated. Indigenous people relished the taste of them and believed their feathers and bills held curative powers; curious shooters killed them as a novelty; insatiable loggers moved into their vast forest empire and felled many of the great trees across their range—all played a role in the bird's downfall.

Many biologists have already written the Imperial Woodpecker's epitaph, and it's a sad tale of massive habitat destruction and wanton killing. But stories still persist among the mountain villagers of lone Imperial Woodpeckers flying over remote pine forests of the Sierra Madre. And this is what kept me going.

The last documented sighting of an Imperial Woodpecker took place in 1956 in the state of Durango in the high-altitude old-growth pine forest of the Sierra Madre. It was this sighting, by Pennsylvania dentist and amateur ornithologist William Rhein, that drew the attention of my friend and colleague Dutch ornithologist Martjan Lammertink and me and eventually led us to launch an expedition to explore the area where Rhein had filmed a lone female Imperial Woodpecker. Amazingly, the 16 mm color movie footage Rhein shot in 1956 is the only photographic documentation ever captured of a living Imperial Woodpecker. Yet for decades the scientific community knew nothing about it, and that might still be the case if not for Martjan, who had seen a mention of the film while reading through a 1962 letter from Rhein in James Tanner's personal correspondence, archived at Cornell University. Rhein wrote that he had "some very poor footage of a female ivory-billed with several short flight shots taken hand held from the back of a mule." (These birds were often referred to as Imperial "Ivory-billed" Woodpeckers at that time.) Martjan was determined to see the film. It took a long time, but he finally tracked Rhein down at his home in Mechanicsburg, Pennsylvania, and watched the footage. He was stunned by what he saw.

As the film flickered on the portable screen in Rhein's living room, a huge woodpecker suddenly appeared, its forward-curved black crest bouncing jauntily as it hitched up a massive pine trunk. There was no doubt about it—this was an adult female Imperial Woodpecker. The bird foraged, chipping off chunks of bark, and then flew away, but the show didn't end there. Rhein had filmed several other short segments of the bird as she flew, foraged, and hitched up trees—sometimes in real time, other times in slow motion. This single eighty-five-second clip held a gold mine of information, and no one else had ever seen it before. Rhein said he would have a copy of the film made for Martjan, but he didn't receive it until after Rhein passed away a couple of years later.

The first time Martjan showed me the movie, it gave me chills. Here was footage of this near-mythical bird—tangible evidence of its existence taken during my lifetime—and it somehow made the quest seem more real. Although I'd already been planning to go to Mexico to search for imperials, viewing Rhein's film pushed me over the edge. I soon started making forays into the high country of the Sierra Madre.

It was a different world there, like stepping back in time to the nineteenth century. The people—many of whom are Tarahumara, Tepehuan, or other indigenous groups—live mostly in adobe huts and cabins without electricity, plumbing, telephones, or other modern conveniences. They rely on horses and mules—or their own feet—to get around, which is not such a bad thing. The mountain roads are atrocious for any kind of motor vehicle, and in many places it takes hours just to drive thirty or forty miles. But that's not the worst of it. The Sierra Madre has become a major drug-growing region, where illicit crops of opium poppies and marijuana thrive. So in the same places where indigenous people engage in their traditional pursuits you also encounter heavily armed drug traffickers driving gleaming new four-wheel-drive trucks with dark-tinted windows. The situation has deteriorated steadily for several years. Government forces are locked in a deadly struggle with drug cartels, and the level of violence has reached horrendous levels in some areas, with massacres, assassinations, and kidnappings becoming increasingly common. And now it is spreading into some of the most remote sections of the mountains. It's no wonder I have nightmares.

I made four journeys in 2008 and 2009 into the northern section of the Sierra Madre in the state of Chihuahua, traveling through the rim forests along the Barrancas del Cobre (Copper Canyon) and in the high country above the villages of Garcia, Pacheco, and Chuhuichupa, where many Imperial Woodpecker specimens were collected a century or more earlier. I had some unsettling encounters with drug growers as well as heavily armed government troops at remote roadblocks. Often the soldiers had their faces masked by balaclavas to protect their identities, fearing that if drug traffickers found out who they were they might terrorize their families.

Still, I was eager to explore the area surrounding Rhein's filming site with Martjan. This would be my first foray into the state of Durango. We'd been talking about it for months, and the expedition seemed more and more imperative as we studied satellite maps of the area. Using Google

Earth and clues in several of Rhein's letters, we tried to decipher exactly where the footage of the woodpecker had been taken, and while doing so, we noticed several areas of uncut forest nearby on high mesas with steep rocky sides.

In the 1990s, Martjan had found people in the Sierra Madre who'd had seemingly credible sightings of the species, and I had interviewed some men in 2009 who had seen birds within the previous five years that matched the description of an Imperial Woodpecker, so it seemed possible a handful of them might yet linger on. As we sat looking at those Google Earth images on a computer, we became more and more excited about going there. What if these places are tiny relics of a pristine ecosystem where Imperial Woodpeckers might still survive?

Martjan had developed a new way to look for *Campephilus* woodpeckers (the genus that includes Ivory-billed and Imperial Woodpeckers as well as nine other Latin American species), and we were eager to try it out in the mountains of Mexico. He had created a device to mimic their typical double-knock drum, hoping that if an Imperial Woodpecker were in the vicinity it might respond with a double knock of its own. The typical double knock is lightning fast and sounds like *BAM-bam*, the second *bam* fainter, almost like an echo of the first. The device has two parts—one looks almost like a wooden birdhouse and is strapped tightly to a tree trunk to act as a resonator, amplifying the sound and giving it a natural wooden quality; the other consists of two broomstick-sized pieces of dowel, each nearly a yard long, connected to each other with a pivot bolt. You swing the dowel assembly overhand in an arc, striking the box with one of the dowels, and the other dowel swings over and down from the pivot point, hitting the box a fraction of a second later and not quite as hard as the first—*BAM bam*—making a sound identical to that of a *Campephilus* woodpecker. Martjan had already tried it successfully with Pale-billed, Robust, Cream-backed, and Magellanic woodpeckers in Latin America, drawing double-knock responses from each species.

Things started moving quickly as we approached our expedition target date in late February 2010. Martjan knew a state forester in Durango named Julian, who was familiar with the area we wanted to explore and said he would help us with our arrangements and introduce us to some of the local people. And I had earlier located the last surviving member of the 1956 expedition, Richard Heintzelman, who was in his early twenties when he went to Mexico with Rhein and another man, Dick Rauch. Although he didn't remember all the details of the expedition, Heintzelman sent me a wonderful selection of color slides he had taken in Mexico. I made 8 x 10 prints of them to take with us so we could try to locate some of the exact areas they had explored, and compare the habitats then and now. One of the slides showed a distinctive rock outcropping looming over the landscape, with a man in a sombrero leading a couple of mules. Heintzelman told me that if we could find that outcropping, it was very close to where they camped—and where the film was taken.

When we met Julian on our second day in Durango, we played Rhein's film on Martjan's computer and showed him Dick Heintzelman's pictures. He paused at the photograph of the rocky outcropping and said he knew where it was, not far from the village of Guacamayita. Later, when he introduced us to some Tepehuan men from the Sierra Madre visiting Durango, they smiled and said, "Los Pilares" (The Pillars). I should mention that the men were narcotics growers and their boss, a drug lord who I'll call Carlos, was with them. Remarkably, they seemed enthusiastic about our proposed expedition into their territory. Julian had told them what a wonderful thing we were doing, trying to save their native wildlife, and they should help us. Carlos agreed and said he would escort us into the mountains and introduce us to various village elders. It was an incredible relief that an important local person would be helping us—and yet terrifying because he was a drug lord, albeit a kinder, gentler one.

So now we had a plan and knew the exact location of one of the places we hoped to explore. At that point Julian didn't think we'd have trouble traveling there. But he warned us against entering a couple of other areas we had originally hoped to check, because they were now controlled by Los Zetas, one of the most dangerous drug cartels. The group was made up mostly of well-armed paramilitaries who had formerly been elite soldiers in the Mexican military before leaving to reap the huge profits to be made growing and trafficking drugs.

Before we even began our journey to Los Pilares—a five-hour drive away on a terrible dirt road—conditions began to deteriorate, and a wave of violence began sweeping into the area. The evening before we planned to drive into the Sierra Madre, we heard an explosion. Someone had shot a rocket-propelled grenade at a hotel just down the street from us where Federales (Mexican federal police) were staying. And then Julian received an anonymous phone call from someone who knew all about our travel plans and was furious that we were going into the mountains.

Julian asked us to meet him just before midnight at the Plaza de Armas, the central plaza of Durango, not far from our hotel. As we walked along the flagstone streets beside the old cathedral, we noticed the perimeter of the plaza was full of police. Their white squad cars lined the edge of the street, red lights flashing on top. I'm sure they wanted to establish a police presence after the grenade attack earlier. We walked past them without being stopped and strolled into the Plaza de Armas. Somehow the police didn't make us feel any safer. Perhaps they might even draw an attack from someone else, and we might get caught in the cross fire. As we walked along in semi-darkness, people kept emerging from the shadows, filling us with dread. The atmosphere was surreal.

Julian showed up fifteen minutes late and had Carlos with him and a Mexican researcher named Marco who was supposed to go with us. Carlos was young—perhaps in his late twenties—short and thin, with a darkly tanned face and a scar on the upper part of his left cheek from an old wound that looked like it had been stitched up at home. He would smile readily, but his face had a way of turning grim in an instant as his mouth dropped, like a cloud passing over the sun. But tonight he seemed confident—even blasé. He said we could just leave earlier. There shouldn't be a problem. But Julian had a stricken look and said it was a bad idea to go there now. He asked us to reconsider. He apologized for his earlier encouragements and said the situation had changed; the whole area had become far more dangerous in recent days. A couple of Mexican researchers who were supposed to go with us had already dropped out of the expedition over safety concerns.

"Of course, if you want to go, I'll still go with you as we planned," said Julian. "It's up to you."

Martjan and I glanced at each other.

"Do you want to go ahead with this?" Martjan asked. "We could always come back another time."

I paused for a few seconds, imagining Los Pilares in my mind, how close it was—just a single day's rough drive—and yet so terribly far away, across a broad dark chasm of unknowable danger. I wanted to go there so badly I could close my eyes and almost feel the place. I knew in my heart this would be my only chance to go there, and perhaps the only chance that anyone would go there for decades to come. It seemed every time I planned to travel to the Sierra, Mexican friends would warn me not to go right now and to wait until the next year when the situation might be better. But each year seemed to be ten times worse than the year before. What if there *were* Imperial Woodpeckers up there? No one would ever know. It had to be now or never. I drew my breath and took the plunge.

"Yes," I said. "Yes, I do want to do this."

And that was that. I had chosen to risk everything for a bird, albeit a beautiful one perhaps reduced to a handful of individuals if it still existed at all.

Julian's face dropped, but he nodded and said he would help us. At that instant I knew whatever happened would be my fault, and it was a grim realization. I was putting not only my own life at risk but the lives of everyone involved in the expedition.

Marco had been very quiet up to that point but then said: "I'm sorry. I can't do this. It's just too dangerous." He said he was going to drive back to Mazatlan in the morning.

Our core group traveling to the Sierra Madre consisted of Martjan and me and two young field assistants, Oscar and Manuel, from the Mexican conservation group Pronatura Noroeste, which also supplied the four-wheel-drive pickup truck we would be driving. On the first day we drove in a three-vehicle caravan—the men from the Sierra Madre we had met the day before in front; Oscar, Manuel, and me in the second truck; and Martjan and Julian last in a forestry truck.

The drive from Durango to the tiny village of Guacamayita was horrendous—a pounding, hours-long grind through remote mountain pine forests on a dirt track so rough and full of potholes and boulders you could drive only four or five miles an hour in many places. En route, a dark-colored club-cab pickup came racing up behind us and then overtook our truck. As they sped past, I saw the men inside holding automatic weapons. When they caught up with Carlos's truck, ahead of us, they drove beside him for several minutes before tearing off up the road.

When we finally reached Las Espinas—a tiny Tepehuan village in an area Carlos considered safe—he and his friends stopped and got out of the truck. Carlos looked visibly shaken and told us the men in the other truck fit the description of some people who had murdered a man in another village just a week earlier.

An hour later, I saw Carlos standing outside his cabin, loading bullets into a wickedly compact Uzi submachine gun. He smiled as he saw me staring with my mouth agape. "*Por coyotes*," he said, smiling as he opened the door of his truck and slid the weapon carefully under his seat. This was clearly an escalation. Carlos would never again go anywhere unarmed—despite the fact that it was illegal to carry guns in Mexico, and he faced being arrested on a serious charge and having his truck confiscated.

Julian had planned to drive back to Durango the same day, but Carlos told him it was far too dangerous. He should wait until the next day when they could accompany him in some other vehicles. And to us Carlos said we should not under any circumstances attempt to drive out of the mountains unescorted. He and Julian would meet us here in two weeks at ten o'clock in the morning, and we would drive out of the mountains in a three-vehicle convoy, the same way we had come. But even as he said this, I had my doubts. This is the Sierra Madre, after all, where events rarely unfold as planned.

I don't want to dwell on the dangers we faced in the mountains. Suffice it to say that a grim shadow stalked our travels in the Sierra Madre—a sense of uneasiness, especially when we would run across an opium or marijuana patch or see someone casually carrying an AK-47. It felt like anything could happen. While we were in the mountains a man was randomly abducted from the nearby village of Las Espinas and held for a 10,000 peso ransom (about $800 American)—a colossal amount for the local people, who all pitched in to buy the man's freedom. But what would happen the next time kidnappers grabbed someone? Later, we saw the charred rubble of three wooden houses that had been deliberately burned a day earlier—perhaps the grim result of a failed extortion attempt.

During our two weeks in the high country, searching for Imperial Woodpeckers, we rose most days before dawn and hiked until dark. These were grueling, at times horrendous treks, climbing up and down treacherous chasms, through pine forests and streams or along narrow knife-edge ridges and across lofty mesas in the scorching sun. According to our notes, we hiked seventy-three kilometers—which doesn't sound like much until you factor in the altitude (often at or above 9,000 feet) and the ruggedness of the terrain as we scrambled up and down rocky, near-vertical canyon sides on crude trails, or no trails at all.

Stopping at places with a good open view of the forest, Martjan would tie the double-knock box securely to the side of a tree trunk and do a series of double knocks—*BAM-bam*, *BAM-bam*, *BAM-bam*—and pause for twenty minutes as we listened closely for a response, then repeat the process. He also played some Ivory-billed Woodpecker calls recorded in Louisiana during the 1935 Cornell expedition. According to Cornell Lab founder Arthur A. Allen, who had heard the calls of both Ivory-billed and Imperial Woodpeckers, the sounds were virtually indistinguishable, like the toots of a child's toy horn.

We ran through this protocol at eighteen different points in the areas of roadless uncut forest and at five points in the vicinity of where Rhein shot the film of the lone Imperial Woodpecker. These points were spaced at least 500 meters apart and offered clear views of the forest or across valleys where the sound would carry a long distance. We never heard a response.

The area where Rhein made his film had altered drastically since 1956. The old-growth pines on this broad undulating plateau are long gone. The place has been logged repeatedly in the intervening years, and the lush grasses that once thrived beneath them have been grazed away. Most of the trees are small now, growing in dense stands, nothing like the park-like expanses of the old-growth forest. But we did find some beautiful areas of uncut forest—the ones we had been viewing for months on Google Earth. Although the truly massive pines did not grow as readily there, because the soil was dry, rocky, and less rich than in the vast tablelands where the film was shot, there appeared to be enough suitable habitat to support Imperial Woodpeckers. Perhaps additional forces had been at work in the birds' disappearance.

We interviewed several elderly residents who remembered the birds well, though none of them had seen any since the 1950s. One man told us the grim story of a forester in the early 1950s who had encouraged local people to kill Imperial Woodpeckers because they were destroying valuable timber. (I guess it made no difference to him that the birds only went after trees that were infested with beetle grubs.) He even supplied the villagers with poison to smear on the birds' foraging trees. It's easy to imagine these poisoned trees attracting Imperial Woodpeckers from miles around—even from the remote and barely accessible mesas we were exploring—and killing them. Shooting the big woodpeckers also seemed to be rampant. The previous year, I had interviewed a ninety-five-year-old rancher in the mountains of Chihuahua who remembered seeing six dead Imperial Woodpeckers lying in front of a sawmill during the 1950s. And bird artist Frederick Hilton—who accompanied Rhein on his first journey into the Sierra Madre in 1953—was told by some local men that a dozen Imperial Woodpeckers had been killed just the year before. Hilton told me this was completely believable considering the number of large woodpecker cavities he had seen in the area. It seems clear that a large-scale extermination campaign was waged against these birds, similar to what was being done at the same time to the wolves and grizzly bears of the Sierra Madre, both of which are gone now. The population crash of the Imperial Woodpecker was not just collateral damage from logging.

Our expedition was one of the most sobering and depressing journeys of my life. It all seemed so hopeless. Even if we had found Imperial Woodpeckers, I doubt we could have saved them. One afternoon as we stood on the rim of a canyon, looking across to the area controlled by Los Zetas, the forest on the other side was aflame in a blaze deliberately set to create more land for drug cultivation. This had been one of the areas we had originally wanted to explore. As we stood there, Martjan held a Google Earth map of the place, made with satellite images taken in 2005, when it was pristine. Now the trees were going up in a cloud of smoke.

On the final morning of our journey, as I had feared, no one showed up to drive out with us. We spoke with one man in the village who told us he hadn't seen Carlos for days and didn't know when he'd be back. He said there'd been a lot of violence on the road down the mountain. We were stunned but decided to drive to Durango as quickly as possible.

We rode in silence for a time, deep in thought, and nearly gasped when the first vehicle came around a corner ahead of us, a battered old flatbed truck. A bunch of Tepehuanes sat huddled and scared on the back of it like refugees—old men, women, and children wrapped in blankets against the cold. The truck stopped alongside us, and the woman riding in the passenger seat bent across the driver to speak with us. She said the road ahead was dangerous and to be very careful. Then, before they drove off, she said, "God may help you," and crossed herself.

At this point I realized we had probably a fifty-fifty chance of running into trouble on the long drive back to Durango. In some places, you could drive only three or four miles per hour—not as fast as most people can walk. It would be so easy to ambush a vehicle. Just step out from behind a tree, holding an AK-47 to the windshield, and all we could do is stop and let them do as they please.

I decided to remove the memory cards from my cameras and my digital voice recorder and put them in the hip pocket of my jeans. I also stuck the notebook in which I'd taken copious notes during the expedition and my passport in the pocket of my coat. I figured if some garden-variety highwaymen stopped us they might just take the truck and everything in it and leave us on the road, waiting for someone to come along and help us. If that happened, at least I would still have all of my travel documents, notes, and photographs, so my trip would not be a waste.

But there was also a thought deep inside me that things could easily go badly. It was a grim fantasy, and I couldn't get it out of my thoughts as we drove along in silence. I went over my life in my mind, and I felt many regrets. I thought of my wife and our years together, and I was sorry I hadn't told her how much she means to me. And I regretted I wouldn't be able to watch our children grow into adulthood and be there to help them accomplish everything they dreamed of doing in life.

When we finally reached the paved road, our sense of relief was overwhelming. Somehow we'd gotten through without incident, passing only one other truck on the road down from the mountain, and its occupants seemed as frightened as we were. We arrived at our hotel, ragged and filthy, but so grateful to be alive. Julian called later and made arrangements to meet Martjan and me in the Plaza de Armas the following morning. I'll never forget how he looked as he walked across the plaza toward us—so intense, so emotional. He hugged us both tightly and said he would never have forgiven himself if anything had happened to us. He knew about the violence sweeping through this part of the Sierra Madre and wanted to warn us not to try driving out alone, but he had been unable to get a government truck to come and meet us. He was horrified when he heard that Carlos wouldn't be able to go there either. Julian told us that the night before, he and his wife had gotten down on their knees and prayed to God for our safety.

The three of us stood in the plaza together, tears streaming down our faces. I closed my eyes and in my mind I saw an Imperial Woodpecker hitching up the trunk of a massive pine. The bird looked back once over its shoulder, then flew off through the forest. It was a sight I knew I'd never see. My search was over. I had made a decision. Finding an Imperial Woodpecker was not worth losing my life or destroying the happiness and well-being of my family. As the thought entered my mind, I felt a massive weight lift from my spirit. And I never had that nightmare again.

Chapter 22

Extralimital

Alison Világ

In Alaska's Pribilof Islands, spring is chilly and drawn-out, with little fits of winter that persist. For the guides of St. Paul Tour, many days bear some semblance to an endurance feat: overtime in the elements in the land of a midnight sun rarely seen. During the two springs I guided there, we saw Wood Sandpipers—Eurasia's equivalent of America's Lesser Yellowlegs—on more days than we saw blue skies. During spring storms, violent gusts toppled tripods, made guillotines out of the doors of our twelve-passenger vans, and basted eyes, ears, and optics with volcanic grit. You could type "get sand out of" and any of the options Google predicted might complete the sentence would be valuable advice.

But the wind, when it came from the right direction, blew in the good birds—the ones from Russia. Alaska is often romanticized as some semblance of the final frontier. For the serious North American birder, it is. Here, the New World's mainland reaches out towards the Old; the two come as close as they will to touching. There lie little scraps of land that break—if only barely—the water stretching between the two continents, and in the mid-1970s, serious birders began to journey to them. Soon, these islands—especially Attu, the outermost Aleutian, St. Lawrence Island, and St. Paul, in the Pribilofs—became revered vagrant traps: places where one could almost depend on encountering an aggregate of birds virtually never found on North America's mainland, such as Common Snipe and Siberian Rubythroat. It's not even ludicrous for visiting birders to bring with them dreams of discovering a species never before found in North America; in these places, this seems to happen most years. The outpost islands of Alaska are, to the serious North American birder, some of the most exciting places to indulge the craft. Every migration—even every storm—deals a unique hand. The odds are always slim, but the stakes are high. Sometimes, the alchemy of wind, of season, and of being in the right place at the right time works itself out. And imagining this formula in perfection is the driving force of the allure of a place like St. Paul.

St. Paul Tour is founded on the daily work (and daily bread) of finding these out-of-place birds. Most of our guests visit because they are bent on seeing birds. Birds that, ordinarily, should be in Siberia. There is an irony in the dogged intentionality it takes to design a trip on the basis of seeing species described in North American field guides as "accidental." But the birder charges of St. Paul Tour arrive with visions of birds that, by weather or error or reasons unknown, turn up on St. Paul instead of the Kamchatka. These birds come with names evocative of exotic origins: Siberian Accentor. Eurasian Hobby. Asian Brown Flycatcher. They come on a southwest wind, and the responsibility of finding them falls upon St. Paul Tour.

St. Paul Island Tour is one of the business endeavors of Tanadgusix Corporation, the Alaska Native corporation attached to St. Paul's Aleut community. In May, three fortunate birders travel to the island, where they will remain until October, to guide. To be a paid witness of a spring and fall migration in the Pribilofs is a dream job for the itinerant bird bum. In 2015 and 2016, I was one of those lucky three. At the time, I was partway through a degree in environmental writing at a small Maine college, and to me, St. Paul was making good on a lifelong dream to work in Alaska. Though prior to this job I'd never visited the state, a year spent living in the Philippines had given me a fair-enough grasp on Asian birds to be dangerous.

This was not my first guiding gig, either. That happened in northern Michigan's jack pine barrens, when I was eighteen and barely graduated from high school. It was my first real job, and the primary duty was to put Kirtland's Warblers into a spotting scope for the day's entourage of warbler-seekers. At the time, this charismatic slate-and-lemon songbird was endangered, its nesting range restricted to a few counties in northern Michigan. I was based in Grayling, a town that for many people deviates from the beaten path. Many tour participants travel there for the singular reason of seeing a Kirtland's Warbler, but as long as they make that journey, odds are in their favor. If ever there was a bird that wanted to be seen, it is a Kirtland's Warbler on its breeding grounds. In the correct habitat, they are abundant, their song is loud, constant, and they are not shy, oftentimes posting atop a tree for many minutes to sing. In the guiding business, these warblers are sort of an anomaly, at once both highly desirable because of their range and status yet extraordinarily reliable. Never in three years guiding in Grayling did one of my groups fail to see a Kirtland's Warbler.

Unlike the Kirtland's Warbler, the birds that summon people to St. Paul do not want to be seen. These, in contrast, are lost, confused, ill at ease, and dart cryptically between patches of cover. Where birders journey to Grayling with the clear intent of seeing Kirtland's Warblers, birding on St. Paul is far less tangible. Birders arriving on the island do not know what the find of the trip will be. They don't know where it will be. They don't even know *if* it will be. St. Paul Tour is unique among birding tour companies in that its business is, to a large degree, vagrant-oriented. To market variables that are furtive and winged—and very possibly not even present on the island—is high-pressure guiding. You, the guide, are the local expert. The burden of the mission falls on you, the closest thing to a responsible party upon which to attach celebration—or blame.

When the guides arrive on the island, it is early May. Rotting snow, down parkas, and a chill linger on. Ghostly white McKay's Buntings, held over from winter, might still be found picking through the detritus near Marunich, up on the north shore. Most of St. Paul's features are shades of brown: dead grass, shaggy reindeer, Bar-tailed Godwits bathing in a puddle in a red-brown scoria road. The decor at the King Eider Hotel, which is also the airport as well as the power station—in Alaska, it is standard to wear many hats—includes a big world map and a checklist of every bird species ever seen on St. Paul. As the season progresses, and more of the regulars are found, this checklist becomes highlighted in yellow and orange. (And, we hope, birds not yet on this list will be scrawled in the blank space at the end.) The map fills with color as visitors mark their hometowns with a thumbtack. But in this preamble to tour season, there is nothing. The map and list are empty, waiting. It is anyone's guess what the season will bring, what it will hold, what it will lack—how it will surprise, how it will disappoint.

Before the guests begin to arrive, a few days are allotted for reacquainting oneself with St. Paul's quirks. This is the time to search out wayward sets of keys, to confirm that Trident Cannery—the island's sole eatery—is ready to feed a pack of hungry birders. You inventory the vehicles in the tour fleet: twelve-passenger vans that go places most four-by-fours don't have short life expectancies. If the engines aren't knocking, the windows close with minimal engineering, and the brakes work, the season is already off to a good start.

The last night before the first visiting birders were expected on St. Paul—the final night of freedom—I walked to Reef Cliffs, just outside town. The air was a heavy fusion of kelp and seal rookery; Steller's sea lions coughed; seals turned over and over in the froth. Below and to the left, the cliffs spread like some sort of geologic origami; seabirds like murres, auklets, and kittiwakes settled on some of these folds, claiming their stoops, but many were yet to appear. As the season went on, these cliffs, like the checklist at the King Eider Hotel, would fill with birds. I put my face into the wind and wondered what it would bring. Everyone has expectations, desires—the guide, committed to St. Paul until mid-October; the client, who will soon step off the plane. We all want good birds.

In the Pribilofs, migration plays out much differently than it does in places like the Gulf Coast or Magee Marsh. These latter sites fall along the flyways, the major travel corridors birds follow when moving between winter and summer grounds. St. Paul, though, is in the middle of the Bering Sea. It really isn't on the way to anything. Along the flyways, in particular at a formidable water crossing, thousands of birds may stack up, waiting for the weather to turn in their favor. Beholding these migrant masses is something like being at the epicenter of a multisensory kaleidoscope: the leaf litter pulses with the rustling of hundreds of sparrows and thrushes while, overhead, warblers pass by the dozen. Witnessing something like this will bring you to your knees. On St. Paul, chasing migration will bring you to your knees, too—but literally, from tripping over tundra hummocks. In contrast to mainland North America, Priblovian migration at its best is often limited to a steady trickle: a few shorebirds at Town Marsh, or a couple of pipits darting around a lava field. But though in volume this trickle is slow, it can provide some of the most exciting birding in North America—when the wind is right.

When I arrived on St. Paul in 2016 for a second season of guiding, I had high hopes for a quality spring. We did have a stellar start: the first full day the guides were on the island, I found an adult male McKay's Bunting and an adult male Brambling—in the same binocular view! Later that night, I was walking to the bar to meet a friend and was waylaid by a White Wagtail. Naturally, this made me late. Soon after—the next day or so—Asian shorebirds began to filter in: there seemed to be a Wood Sandpiper or two on every little wetland; little flocks of Long-toed Stints scurried around the wet grassy areas; several Common Sandpipers teetered along the rugged shorelines. There were Common Snipe and a Curlew Sandpiper, and three very hopeful guides.

We still had a couple days, though, before anyone else would be there to share in the excitement.

On our final day without clients, my coworker Claudia and I were driving back towards town—and lunch—after a morning spent trying to kick up vagrants. Before we made it to our abode (and its awaiting spanakopita), my phone rang. It was Scott, the tour director and third guide—he was looking at a Long-billed Murrelet in the town's bay! This was Scott's seventh year working on St. Paul. The murrelet, old-world counterpart to our Marbled Murrelet, was new for him. It was a *good bird*. Speeding over, we passed by Salt Lagoon and noticed a lanky shorebird on the near edge. We backed up into our dust cloud to quickly snap a record shot of the season's first Common Greenshank before continuing on to the murrelet.

But then, the clients started showing up, and, frustratingly, the winds shifted back east, where they would stay for most the rest of the spring. By summer's solstice, after more than a month of squinting over wind maps, of tracing their colorful squiggles leaving Asia (and losing them somewhere over the Okhotsk or Bering long before they ever reached St. Paul)—a month of willing winds from all the wrong directions to change—we had acquiesced a little. Of course, there was still hope for a late-season migrant, like a flycatcher or a cuckoo; ungrounded optimism is the best attitude to have when birding St. Paul. The serious birders, however, were phasing out, replaced by equally serious photographers coming to capture the charisma of the cliff-nesters—Crested Auklets and Red-legged Kittiwakes and Tufted Puffins. Earlier in the season, we had used the long subarctic days to comb boulder fields and slog marshes. Now though, much of our time was spent at cliffs or traveling between them. After the high-traffic, high-intensity spring migration crowd, this was an appreciated respite.

On a day in the last week of June, I was guiding four people, the only clients on the island at that moment: John and Janet, Brit photographers; and Sam and Gail, birders. Scott and Claudia had the day off, and it was a good day to *have* off. Sometimes, a group of mixed clients breeds discord—the birders want to scour a slew of sites for vagrants, while the photographers want to sit in one spot for an entire morning and take a slew of photos. However, on this morning, there was no such friction: the five of us were united in a common desire to be inside the van. Priblovian weather had driven us off Reef Cliffs early; it was the sort of day where you spend more time reclaiming your optics from sea spray than you do actually looking through them. We were traveling towards Ridge Wall. This was not because I expected conditions to improve—all of St. Paul's cliffs are a rough time in southeast gales—but because traveling to Ridge would allow us the longest recovery period before exiting the van to certain cold and misery somewhere else. My body was sore from shivering and already I was anticipating the evening ritual of savoring a cup of Bailey's and hot chocolate from the staff house boiler room.

Just outside town, we stopped at Salt Lagoon, an expansive tidal flat that sometimes holds an interesting shorebird or gull. I peered between the raindrops on my binoculars, noting that not even the condensation enhanced the appearance of a Glaucous-winged Gull: they all still looked various cycles of sorry. I weighed the options. Guides on St. Paul, like workers anywhere, carry a toolbox. Not just physical—lens cloths and spotting scopes and *Rare Birds of North America*—but one less tangible. Tools like a soft touch to ease a decrepit twelve-passenger van out of a treacherous sand pit; the sense of working with wind, light, and a Jack Snipe so the group gets the best views as it flushes; how to best console a client whose luggage fell victim to the wiles of PenAir.

Almost since the season had started, "endurance," "optimism," and "the art of convincingly using time between Trident meals" had been tools in regular rotation. The fun tools—"pin a name on that Asian flycatcher," for example—had not. The winds just had not been with us. We all, guides and clients alike, had been poised and ready, begging the winds for an opportunity to use our full repertoire. There was so much that we could do. But, there also was *only* so much we could do, for we needed luck, and it was beginning to seem that luck was largely absent this season.

On this day, though we did not recognize it at the time, it was luck that the bad weather had driven us from the cliffs early. It was luck that Gail asked for a scope view of the Red-legged Kittiwakes resting on Salt Lagoon's tidal flats. It was luck that, after setting her up on the kittiwake flock, I happened to glance skyward at the right place, the right time. A bird cut over the harbor towards us. Something about it seemed different. I was catching a weird angle, perhaps—maybe it was a *Tringa* of some sort? A Wood Sandpiper would always be welcome. I swung my binoculars up and immediately realized my error: the long projection that, naked-eye, I had assumed to be legs was actually a tail. It was a swift—one much too large for the expected American species.

"*SWIFT!!!*" I shrieked. "GET ON IT! IT'S ASIAN!"

Suddenly, those tools I had been aching to use all migration came into play. I dropped my bins, raised my camera, and fired off shots, as many as I could, before the dark manifestation of aerodynamicity could vanish into anonymity. *Dark*. That had been the first impression we'd all gotten.

"Well, whatever it was, it was an ABA-bird for me," said Sam, speaking for us all.

Swifts are aptly named; this one was gone, evidenced only by my adrenaline shakes and our photos. We bent over the cameras to dissect the identification. There were three possibilities: Pacific Swift, White-throated Needletail, and Common Swift. The first shot showed the underside of the bird. This was black, which meant our bird was not a needletail. I scrolled back through the camera. Thankfully, there was a dorsal view as well. This too was dark—the swift was no Pacific, the most expected of the three.

"All dark all right! It was a COMMON!"

Despite its name, Common Swift is anything but—in North America. In fact, at the time, the number of occurrences of this species in North

America could be counted on one hand. This all is according to *Rare Birds of North America,* the only North American field guide where you'll find Common Swift. In flipping through the pages of *Rare Birds,* you'll find a lot of "common": Redshank. Rosefinch. Cuckoo. Sandpiper . . .

So what are they? Common? Rare? That depends on where you are and where you care. If you're someone like Sam, who has birded on Attu and devoted much time to chasing down birds inside the ABA-area, Common Swift is a mega. On the other hand, if you identify more with John and Janet, the photographers, a Common Swift is fittingly named.

"We've got them in our garden back home," they laughed.

Where is St. Paul, really? Any map would show it being far from *anywhere.* The Bering Sea Islands, of which St. Paul is one, are a fringe spot where borders blur. A cursory glance at a map may not reveal which flag they fly; at 200 miles from Alaska's mainland and 500 from Russia's, there is nothing nearby. A map from 2016 would designate the Pribilofs to be property of the United States of America, but a map from the 1830s would have them as Russia. In town is a Russian Orthodox church whose congregation bears Russian surnames; there, Russian—and Aleut and English—are spoken. Fourth of July, a decidedly American holiday, is celebrated in decidedly American ways: beer, a loud parade, and a community barbecue, it all trimmed in red, white, and blue.

The birds on St. Paul are a conglomerate, too: the *crecca* subspecies of Green-winged Teal reaches the eastern extent of its range; the *carolinensis* subspecies the western. Here, they meet—and, as it will go on small islands, mingle—producing offspring that, like the community, portray the New and Old Worlds alike.

But to an ABA-area lister, it's pretty clear. St. Paul is definitively (albeit barely) inside the ABA-area: a volcano made island in a man-made construct. Listers tend towards the limits of this regulation area—southeast Arizona and Newfoundland and the Florida Keys. At places like these, birds extralimital show up from time to time.

"Extralimital," a visiting birder had mused earlier that spring. "I love the word. It conjures being in the right location in the right place, at the right time, when the alchemy of wind and El Niño and season finally works itself out."

This day, finally, after a season of winds from the wrong directions, the alchemy had come through. Though the swift was long gone, we had some time yet before Trident would have lunch out. It was back to the guide's toolbox—back to the art of convincingly using time. We continued on to our original destination, Ridge Wall, where we would watch the Crested Auklets and Red-legged Kittiwakes and Red-faced Cormorants. These species, to the international birder, constitute the marquee birds of the Pribilofs: they are birds absent from most of the world yet abundant on St. Paul. What is common—what is rare?

Chapter 23

Death and the Rose-breasted Grosbeak

John R. Nelson

In a ten-month span, my only brother and three remaining sisters all died. Within a few years before that, I'd lost close friends, two nephews, and a stepson. Time was rushing ahead. Who's next? Some deaths were more or less expected. Since I'm the youngest by far of seven children, it's not surprising I would outlive my siblings. My oldest sisters, Rosie and Carlotta, were both in their upper eighties when they died. The sudden death of my stepson Tommy, at age thirty three, was a blow. My wife Mary and I were still reeling. For the living, no two deaths are alike. Each is interlaced with a weave of sweet and painful memories, dreams realized or not, twisting patterns of love, lost promise, strands of guilt, empty days, sorrowful yearnings, circling thoughts, and hopes that we can somehow commune with those who are gone.

I also birded a lot that year—for the Massachusetts Breeding Bird Atlas, at Mt. Auburn Cemetery, and travelling around and out of the country in pursuit of new birds. When Mary and I reached Tennessee to find my sister Joan much sicker than we'd known—she died weeks later—I was still abuzz over a bird from a pelagic trip out of North Carolina, Swinhoe's Storm-Petrel, the first North American record. When Rosie died I was in full bird-search mode in the Philippines. When Ralph died, I scrambled to catch a flight to Windsor, Ontario, attended wakes and a funeral, delivered a eulogy, went to an Irish pub with his son, caught a flight home at dawn, and spent the day finding birds for our Bird-a-thon team. I was glad to have something to do with myself.

I usually didn't think about death while I was birding. Near home, on foot or bicycle, I moved through a familiar scape of sight and sound, and potential sound, with near constant concentration. If I were someplace where I didn't know the songs—a tributary of the Amazon—I was still listening and at full visual attention for most of each day and into the nights. Edward O. Wilson, in *Biophilia*, calls it the "naturalist's trance."[1] Gretel Ehrlich writes in *The Solace of Open Spaces*: "Animals hold us to what is present."[2] Iris Murdoch describes a day when she was feeling anxious and resentful until she became entranced by a hovering kestrel: "The brooding self with its hurt vanity has disappeared. There is nothing now but kestrel."[3]

Birding blessedly distracts. Other pastimes, I guess, could serve the purpose, like the tennis games I used to play, or online poker, or Big Days at yard sales. But I never think of birding as a distraction. A distraction couldn't offer such moments of discovery—a Capuchinbird's shocking air raid siren song in a remote Brazilian forest or, at home, a baby hawk flapping and hopping with ungainly resolve and rising in first flight. That spring I spent hours in our yard watching birds' family lives. A pair of Pine Siskins, uncommon breeders in our county, bred in the woods beyond our pond: first the male and female daily at our thistle feeder, then the male in courtship antics—its call a joyriding *zhreee*—then the female carrying grass to build a nest, and, weeks later, the female stuffing food into a fledgling's gape. When a hailstorm hit, I fretted over birds still incubating, battered by hailstones as they sheltered their eggs.

Distractions don't always work. Sometimes grief overtook me, no matter what birds were around, whether siskins at home or a wondrous blue Celestial Monarch flitting through a canopy on Mindanao. I didn't break down sobbing. I was immobilized, disoriented, the last sibling standing, lost. Or I was possessed by a restlessness, uneasy, searching, for what I'm not sure. "No one ever told me," wrote C. S. Lewis after his wife died of cancer, "that grief felt so like fear."[4]

Sometimes a bird set me off. My loved ones weren't birders, but certain birds, to this day, remind me of each of them. With my friend Steve it's the Carolina Wren and Northern Mockingbird, for both sang the whole time—the mockingbird echoing the wren's *teakettle teakettle teakettle*—as we circled his gravestone on a frosty morning. With Joan—beautiful, spirited, elegant in hip waders, a woman who reached out to people—it's any kingfisher, for she could fish expertly whether on a rocking ocean or in a fly-casting stream. With Ralph it's the Canada Warbler. The words for his eulogy came easily. My brother, a political science professor, was a kind, smart, witty, and faithful man. In my book *Cultivating Judgment*, on how to teach critical thinking skills, I'd used him as a model to emulate.

Rosie is brought back by Dark-eyed Juncos, birds I was surprised to see in July on our hike up Roan Mountain in Tennessee, months after they'd left our yard. I'd forgotten that juncos are altitudinal migrants. Rosie liked their energy, their blithe acceptance of cold and wind, the white flashing in their tails. For her eulogy I wrote a poem, "Woman of a Garden," comparing her tending of her garden to her loving cultivation of her eleven children. For years she'd helped to preserve birds as a volunteer at a nature center, and at the Field Museum of Natural History in Chicago she'd led groups of children and elders to see the bird specimens. Carlotta appears with a flock of terns that called out and streamed over as we strolled along a breakwater in Provincetown, before disease began to claim her. Which species I can't say, since I wasn't a birder then, but it doesn't matter. "My, aren't they grand," she said. Carlotta, a sensitive heart, a nurturing teacher, had a feeling for beauty in all its forms.

I like to think of my stepson Tommy as a closet or incipient birder. We never went birding together, but Mary trained him to look out for Red-tailed Hawks, and he'd indulge me with a smile and show of interest when I went on about birds. He loved creatures generally and had adopted an ailing macaw from the animal shelter where he worked. One day he called me. He sounded excited. He'd been out in the woods and had seen some really cool birds. What were they? He described field marks. His attention to detail amazed me. I can't remember all the species now, but one was unmistakably a Rose-breasted Grosbeak. We made plans to go birding together the next morning. When I got off the phone I proudly told Mary, "I've turned Tommy into a birder." She nodded, then burst out laughing. What? "He didn't see a grosbeak," she said. He hadn't been in the woods at all. With some coaching from her he'd been putting me on. I was touched nonetheless. He'd cared enough to play out his role expertly and string me along.

I didn't think of my loved ones every time I saw these birds. Joan liked goldfinches, just as she hated the turkeys invading her yard—almost fifty on one day—but with all the finches at our feeders and the female turkey that visited daily, pausing to defecate on the flat rocks around our pond, I'd have been grief-struck all day long if every goldfinch or turkey prompted thoughts of her. Rather, birds and my departed had taken up residence together in a dense, tangled, unconscious habitat of feeling and memory. Through birds the fullness of love, of grief, became fuller.

One morning before dawn I lay in bed, unable to get back to sleep. From the stillness outside came a song, slurred and wavering, a Veery, a thrush I'd never heard in our woods. I remembered Edna St. Vincent Millay's "This Is Mine, and I Can Hold It." A beautiful thrush song makes her feel like a vessel about to shatter, for all her senses "have broken their dikes and flooded into one, the sense of hearing."[5]

The Veery stopped singing and then started up again. I thought of another man lying in darkness and listening to a thrush. I'd first taught Keats's "Ode to a Nightingale" decades earlier, long before I knew about birds or felt ringed by death. Keats wasn't a birder in the modern sense, skilled at identifying species, but he had that rare quality birders strive to achieve, the ability to stay attuned to the slightest movement, sound, or shift in light—the naturalist's trance. "Nothing seemed to escape him," wrote Joseph Severn, his friend and eventual nurse, "the song of a bird and the undernote of response from covert and hedge, the rustle of some animal, the changing of the green and brown lights and furtive shadows, the motions of the wind—just how it took certain tall flowers and plants—and the wayfaring of the clouds."[6] Keats knew how it felt to be circled by death. His father died when he was eight; his mother, when he was fourteen. For months, as his brother Tom wasted away from tuberculosis, Keats stayed by his side, nursing him and becoming so distraught, so exhausted, he could barely write letters, much less the poems on which he'd focused all his ambition. "I have never known any unalloy'd Happiness for many days together," he wrote his betrothed, Fanny Brawne. "The death or sickness of someone has always spoilt my hours."[7]

Tom died when Keats was twenty-three, not long before he wrote the "Ode." In the poem Keats never mentions his brother by name, but the youth who "grows pale, and spectre-thin, and dies" is certainly Tom. The ode is set on a spring night. The poet's heart aches as he listens to a nightingale sing in an ecstasy. The song stirs intense yearning. He longs to escape with the bird and "leave the world unseen." He wants to forget what the nightingale will never know—the "weariness, the fever, and the fret" of life, the palsy of the old, the sickness that steals the young. Knowing he can't forget, he imagines himself dying, that very moment, a painless death as "easeful" as the nightingale's song. He envies the bird's freedom, not from death, for the bird will die, but from awareness of death. But eventually the nightingale flies off. The music is "fled." The poet is left, as we all are, with his "sole self." The ode ends with a question: "Was it a vision, or a waking dream?" In a sense Keats has flown with the bird: he has transformed its song into a poem. In a world of death, two consolations, if that's the word, have come together, the evocative beauty of birdsong and the power of human language to express the feelings the song evokes. Yet his helplessness, and ours, to fend off death remains.[8]

Keats seemed to know that his own time was brief. He composed the "Ode" shortly before he contracted tuberculosis while on a walking tour in northern England. In early 1819 he wrote that he wanted to "moult" like a bird into a fresh body. By July he no longer "hoped for fresh feathers and wings: they are gone, and in their stead I hope to have a pair of patient sublunary legs." He was living what he called a "posthumous life" when the ode was published the next summer.[9] He died the following February, before his twenty-sixth birthday. When doctors opened his body, the cells of his lungs were so destroyed they could not understand how he'd managed to stay alive the last few months.

The music of my Veery fled, or I fell back to sleep. When I got up, Turkey Vultures were riding thermals over our house. Each year they seemed to arrive earlier and stay longer. A few now winter here. In earlier ages their unseasonable appearance would have been an ill omen, a warning that death was close. In his poem "Under the Vulture-Tree" David Bottoms tries to give vultures a more benign spin. Hundreds of vultures, perched by a river, become "transfiguring angels" with "surprisingly soft faces, like the faces of the very old who have grown to empathize with everything," and "with mercy enough to consume us all and give us wings."[10]

If you've lost a son or brother, or fear that your own time is near, the mercy of carrion-feeders may not provide much comfort, but the need for relief can make us desperate. What do we do with our minds, Keats asks, when our hearts are plagued by sorrow? How can we fly when the "dull brain perplexes and retards"?[11] How can we stop thinking about what we've lost and know we will lose? In her poems about birds and death, Mary Oliver asks the same questions. In "The Return" she asks: what can offer us "deliverance from Time?" What can we do with the knowledge that "death is so everywhere and so entire"? "*What good is hoping? . . . What good is trying? . . . What good is remembering?*"[12] Answers to such questions are a lot to ask of a bird, whether thrush, vulture, or the herons and hummingbirds in Oliver's poems. It's not enough to be distracted or to yearn to fly and forget. We want to be transformed. We want birds to give us faith that we too shall rise. In "What the Thrush Said" a bird says to Keats: "O fret not after knowledge. I have none."[13]

With thrush and vultures gone, I ate breakfast. At our window feeder a bird joined me, feeding nervously on sunflower seeds. I studied it: big pink bill, oversized head with bold stripes, buffy streaking on its breast. It was an immature Black-headed Grosbeak, a species so rare in our state that once I got the word out, birders would be trooping to our yard to see it. Then I noticed the faint, splotchy red wash on its breast. The Latin name for the genus, *Pheucticus*, means "painted with cosmetics." It wasn't a rare bird after all, not a Black-headed but a far more common Rose-breasted Grosbeak, virtually identical as a juvenile except for that slightly blushing breast, which turns blood-red on an adult male. Still, I was pleased. I'd heard grosbeaks singing in the neighborhood and known a pair must be nesting nearby, but I'd never found the required proof—adults building a nest or feeding young—to confirm their breeding status for the atlas. Here was proof indeed.

For a while, in a trance, I watched the bird. It was a scruffy little thing, eyes darting, unsure of itself, probably just fledged. If it were human, it would have acne and a cracked adolescent voice. But in a few weeks it would gather itself and fly, alone, across plains and seas, to a land it had never seen, the Cayman Islands or the Chiriqui highlands of Panama. The next spring it might find its way back to my patch. I thought of Tommy, that sly smile when I let him know I was on to his joke. Yes, you had me fooled, boy. You didn't see a Rose-breasted Grosbeak. You were just having some fun with your old man. How often I regretted the journeys he'd never take, the adventures he'd miss because his life ended too soon, too soon. How seldom I remembered the journeys he did take, to China to perform with his karate group, to Caribbean islands with the girls who could never resist him. He struggled so hard to believe in himself, find himself, lose himself—drugs, too many girls—but he knew joy, and in moments he was beautiful, stroking the sick macaw on his shoulder or, master of self-defense, poised on one leg with arms outstretched like a crane about to lift off. At Mt. Auburn, by an Irish cross with a spread-winged bird of prey and a tree stump with climbing vegetation, all in stone, I'd shivered with a mother's anguish as I stood over a stone crib for a dead infant.

The grosbeak leaped to a perch near the feeder. I got up for a closer look. The bird detected movement but didn't take off. The moment felt intimate, light. I asked the bird no questions, expected no revelation. I didn't envy it, didn't ache for wings to fly south with it. Mortality rates are high for young birds, whatever the cause on the death certificate, hurricane, cat, juvenile pilot error. One day, maybe soon, this bird wouldn't make it. But until that time came, it would be a creature of vitality and beauty, ever striving. I remembered another Keats poem, "Ode on Melancholy." Grief can't be escaped. You can't silence "the wakeful anguish of the soul." No, you must yield, let it come, all of it, with eyes and heart open, and "glut thy sorrow."[14] Wish for no grief, no fear, and you wish for no life at all. I didn't want to flee my dead. I wanted them with me—son, brother, sisters, friends—with birds singing. I glutted my sorrow.

The grosbeak flew off. I heard other birds nearby, a goldfinch, a Carolina Wren's bright *teakettle*, something stirring in a bush. Ridden by loss, Keats still listened for all the signs of life around him. I embraced the day, or tried to.

Chapter 24

Birding on Bleaker Island

Rachel Dickinson

The Falklands—a British overseas territory—is a windswept archipelago set 300 miles off the coast of Argentina in the frigid waters of the South Atlantic. Next stop is South Georgia Island where Shackleton ended his famous attempt to be the first to cross Antarctica. The Falklands is a you-can't-get-there-from-here kind of place, and if you're traveling from North America you have to take the weekly flight from Santiago, Chile. But that means you have at least a week to explore the islands once you're there.

The Falkland Islanders tend to be more British than the British in some ways, and it's like stepping back to the Britain of the first half of the twentieth century as you travel through a land populated by the descendants of Scottish sheep farmers and English sailors. They know how to pull a good pint and serve proper tea.

Three thousand people live in the Falklands along with about a million sheep and ten million birds (I'm making those last numbers up, but you get the drift). People travel from island to island by FIGAS, the island airline service, which flies bright red planes that hold eight to ten people. These planes land on airstrips that are nothing more than grassy pastures, often inhabited by sheep. If you're lucky, the pilot will let you sit in the copilot's seat, and you can chat as you fly over islands and mountains and rough seas.

In 1982, Argentine forces invaded the islands and left seventy-four days later after being thumped by the British. In America, the conflict was portrayed as a silly little war in a god-forsaken place. I was going to the Falklands to research an article about the thirtieth anniversary of the Falklands War. I read about the war before embarking on my trip and, of course, found that like all wars, it was brutal and the politics were complicated, so I knew I was walking into a potentially good story. I was also keen to see some of the birds of the Falklands while I was there. If I was lucky I'd be able to see birds that were new to me, including the Black-browed Albatross, five species of penguin, and flightless steamer ducks. On my first island-hopping trip, I grabbed the little red plane to Bleaker Island, which would get me at least two penguin species, the steamer duck, and maybe a Black-necked Swan.

The owners of Bleaker—Mike and Phyl Rendell—met me at the grassy airstrip and drove me the length of the island (three miles) to the "settlement" (their home, two guesthouses, and the house of the farm manager). There are no roads on Bleaker—there are no roads on most of the islands in the Falklands—so Mike headed overland on a vaguely defined two-track path pointing out where the penguin colonies were and the freshwater pond where I might see a Black-necked Swan. We drove through low-slung patches of diddle-dee, which looks just like heather, and through close-cropped grass on a headland that overlooked a blue and gray sea dotted with distant islands. There are no trees on Bleaker—which is true throughout most of the Falklands—and a large rocky hill covers about half of the small island. Because I hail from the land of trees in central New York State, the sheer openness of the landscape felt raw and exposed.

I had just come from Stanley, the capital city (pop. 2,000), where I spent two days being guided through battlefields by two veterans of the war. These men could recount every single moment of these battles and spoke with an intensity reserved for the battle-scarred. I saw rusting Argentine artillery guns still in place on a rocky ridge and a trail of small white crosses adorned with red paper poppies where British soldiers had fallen during an assault. A curious Falkland Thrush, a bird much like our American Robin, accompanied us up and along the ridge, hopping along the rocky outcrop. I kept thinking about the juxtaposition of the rusting guns and this tame bird; of death and life and of the coldness of the metal and the warmth of the bird. My guide showed me a hole where the Argentines—most of them very young conscripts—had dug in and waited for weeks for the British to come near. He reached behind a rock and held up a decaying cloth boot with a rubber sole that had belonged to an Argentine soldier. The guide placed it back in its hiding place so no one would take it.

No battles had taken place on Bleaker, so it felt like a respite from the intensity of visiting sites associated with the war. This island was where I was going to relax and take a long walk. After getting settled, I put on my boots and rain jacket (weather in the Falklands is synonymous with changeable), slung my binoculars around my neck, grabbed my camera, and headed out to find some birds.

Have I mentioned the skuas? These are huge, predatory birds that look like ubergulls. They are the bird bullies of the islands—harassing other birds to drop their food, attacking and devouring young birds, and swooping and diving on anything they don't like, including people. I first met the very similar Great Skua on Fair Isle in the Outer Hebrides when I was hit on the head from behind by a skua as I walked along a road. I had just passed a tern colony where I saw Great Skuas dive-bombing adult terns that were sitting on their nests. Maybe the skua that hit me didn't like me pausing to take a photo of the harassment. What the photo can't show is the pain of being hit, the stars I saw, and that I will never forget the incident. Now, while walking away from my guesthouse on Bleaker, I saw that there were skuas everywhere. *Oh my God*, why hadn't I seen them on the drive across the island? Mike pointed out a Striated Caracara, a big hawk that reminded me of a Harris's Hawk, and said that they were very curious about people and not to be surprised if they came near. But he never mentioned the skuas that were now patrolling the island. Figuring I needed some kind of strategy to steer clear of the big birds, I walked along the fence line hoping the skuas might take me for a moving fence post. They eyed me, trying to decide whether I was worth harassing.

I climbed through the fence, entering an area of amazing tussac grass—huge five-foot-tall clumps of grass growing out of foot-high root clumps that you had to wedge your feet between to move through. I headed toward the sea. The strong smell of ammonia and a cross between braying and honking hit me as I approached the headland. Rockhopper Penguins! These foot-and-a-half-tall penguins have long, droopy yellow eyebrow feathers, orange beaks, and red eyes, and when they walk they waddle and hop (hence the name). They also have extremely sharp claws that give them purchase on cliff faces. I stood and looked at the messy, loud, smelly penguin colony filled with half-grown birds and molting adults. Some of the rockhoppers looked to be in a serious state of undress as new feathers grew in to replace molted ones.

My next stop was the sandy beach where the Gentoo Penguins come in after a day of fishing. I headed back through the tussac grass aware of how easy it would be to break an ankle among the root clumps. I wondered how long it would take my hosts to find me if that happened. Would Rockhopper Penguins surround me in the tussac grass? Would they peck me with their sharp orange beaks? I shook those thoughts from my head and climbed back through the fence and walked on grass close-cropped by both sheep and Upland Geese. I saw big patches of diddle-dee, and it reminded me of Scotland, where I had lived for a year. I stooped to look at the little pink diddle-dee flowers and I was transported in my mind back thirty years to the Highlands and a day of walking on heather-covered hills. Suddenly, just six feet away a skua rose out of the diddle-dee and came right at me. I ducked and then started to run toward the beach flailing my arms, knowing I was channeling Suzanne Pleshette in the schoolyard scene from Hitchcock's film *The Birds*. For some reason the skua stopped buzzing me when I got to the beach—perhaps it knew it would have another chance to harass me when I headed back to the guesthouse.

I sat panting on the beach from my skua-induced run and watched penguins emerge from the blue-green surf, stand up, then waddle-run toward shore with flippers akimbo. Gentoos are medium-sized penguins, and everything they do seems adorable. They are the *Happy Feet* penguins. And they were curious about me—coming closer and closer as I sat still. The sun came out from behind a bank of clouds, and looking at the Caribbean-colored water, the reflecting white sand, and watching the antics of the bright-white-and-black penguins put me in a relaxed trance-like state until I knew I had to walk back.

I stuck to the beach for as long as I could, searching for a piece of driftwood in an island group not known for trees. I found a three-foot-long piece of wood—what might have been a two-by-four a hundred years ago—and picked it up to use as my skua protection. My plan was to hold it up over my head so the skua would go after the driftwood instead of me. I headed inland trying to stay clear of the diddle-dee, spying plovers and dotterels in the grass on the way to the freshwater pond to look at the Black-necked Swans. But I wandered off course and saw several skuas flying low, looking for something to harass, menace, or eat. I started running, waving the two-by-four above my head while yelling, "Get out of here! Leave me alone!" Two of them double-teamed me, coming straight at me, hitting the stick.

At that point I lost all interest in birds except for the ones that I was convinced were trying to kill me and feed me piece by little bloody piece to their young. The faint two-track road running down the middle of the island lay ahead, and I ran for it. The skuas got bored—they had me right where they wanted me—and left. I held my driftwood high, ready for combat, knowing there could be a sneak attack from the rear.

I passed the tussac grass and the rockhopper colony, spying a pretty little Tussac Wren flitting through the waving grasses. I paused to watch as thousands of large black-and-white Imperial Cormorants flew toward the island, bringing food to their young. As they passed over me like a squadron of planes, I felt like I could reach up and touch them because they were flying so low. They landed about 200 yards away in a colony on the crest of a little hill. There were tens of thousands of cormorants in the colony—many of them young birds just leaving the messy ground nests—and the calls and caws and mewling sounds were deafening as parents sought out their chicks. Several dozen skua glided low over the crowded colony hoping to nab a young bird or a bit of fish dropped by an adult.

I swung wide, again hugging the fence line, and was thrilled when I saw the little settlement up ahead. As I walked between the Rendells' home and the guesthouse where I was staying, I stopped and knelt down to take a photo. Two skua rose from a patch of diddle-dee about three yards away and hovered right in front of me, like they were curious. I dropped my camera and started waving my stick. They looked puzzled then settled back down in the low shrub, watching me as I ran to my guesthouse.

Later, over a wonderful dinner of lamb, broccoli, new potatoes, apple crumble, and a bottle of merlot, Phyl and Mike told me that Elaine, one of the farm managers, feeds the two skua near the house, so they associate people with food. "Why would she feed them?" I asked, and the Rendells told me that it was something she started doing when they were young birds, and now they just hang around. "They wouldn't hurt you," said Phyl. Based on my recent experience, I was skeptical.

We talked about people, politics, and the 1982 war. Neither of them lived in the Falklands when the Argentines occupied the islands thirty years earlier— Phyl was in Britain at school, and Mike, who was from Britain, came to the Falklands as a serviceman right after the war. In a way I was relieved because it meant I didn't have to talk about the war, which seemed to have left those who had lived in occupied Falklands with a kind of community-wide PTSD. The war stories I had chased down while in Stanley were intense—like those from the civilian who helped move Argentine soldiers buried in a mass grave, and the man whose house was bombed and three of his friends were killed. Those stories were told as if the events happened just yesterday and not three decades earlier. Plus, I had been engaged in my own afternoon-long battle with the skua when all I wanted was to take a walk and shake some of the war-story tension from my limbs. Encountering Elaine's pet skua put my jangled nerves over the edge. So the Rendells and I relaxed and talked about birds, birds, birds, and then local politics and books and movies as we listened to the braying and fussing of the Magellanic Penguins in the colony right behind their house. We talked long into the night and watched as the sun set on Bleaker Island and was replaced by a billion stars.

Chapter 25

NEST WATCHER

SUSAN CERULEAN

Some years ago, I began to watch over wild birds along the north Florida coast. I was a volunteer steward, and my first assignment was on a bit of sand, a spoil island south of the Apalachicola bridge. There, I was to keep track of nesting activity by Least Terns, Black Skimmers, certain small plovers, or American Oystercatchers.

Two or three times a month between March and August, I'd travel to Apalachicola and slide my kayak down the concrete ramp at Ten Foot Hole. I'd tidy my lines, floating in the backwater basin between a double row of houseboats, sailboats, party boats, deep-sea fishing craft. I came to know those boats like I did each home on my own street in town. Some of the sailors and anglers recognized me, too. I liked to imagine I was becoming part of the place, the background, not on a first-name basis, but worth a nod, not a startle. The fisher folk might not have known where I was going or what my job was, but I felt as if I were a moving piece of the working waterfront. Not a tourist.

I'd line up the prow of my boat with the red channel markers, and adjust to the tide and the river's wide current. Then I'd push my shoulders into my double-bladed paddle, and align with a mindset that might make me as much a part of the scenery on my upcoming bird survey as I was among the people of the boat ramp. That was my goal. Otherwise, as I entered the birds' nesting ground, I would be perceived as a threat. I imagined fading into quiet, being benign. *I am a simple being, only passing through. I have a familiar aspect, and trajectory. Don't be afraid, not of me.* I cloaked myself in that mantra.

My little island, unnatural as it was, had historically hosted a seasonal congregation of 700 nesting pairs of Brown Pelicans. But after a large quantity of spoil was dredged from the river channel and heaped onto the island one winter, the pelicans abandoned the site and had never returned. Another year, I was told, more than 200 Least Terns and a handful of Gull-billed Terns had nested on the fresh spoil. It wasn't clear whether pelicans would return, or the terns—or neither. It could be a very exciting—or very boring—site to survey and protect.

On my first trip to the island, before my kayak nudged against the hip of the sand, I looked to see if any boaters might be trespassing. None today; that was good. What about aerial predators: Fish Crows, Laughing Gulls or the last of the winter Eagles? An Osprey perched on a post drying its wings, and two Fish Crows stood on the wrack line of the tide. In my new role as steward, I thought of those birds as thugs. All three startled away as I approached.

I rolled up the legs of my pants and slowly circled the island with my spotting scope over my shoulder and my binoculars around my neck, keeping count of all the birds I saw. I soon learned why no other birders wanted to survey this territory. Neither its shape, its smell, nor its vegetative composition appealed to the eye—or the nose. Along the shoreline, my feet sank into oozing mud where hard sand should be packed, were this a true beach. Still, the spoil island had worked as a home for some beach-nesting shorebirds, because it was inaccessible to coyotes, cats, raccoons and was unappealing to most humans.

At either end of the land, south and north, wind drove shallow water against the sand, fussing the wavelets into white-tipped fringe. Birds loafed about in good numbers, bathing and cooling their bodies. Caspian,

Royal and Sandwich Terns stood belly-deep at the island's tips, flapping their wings and shunting water over their backs. It appeared as if the ends of the islands were trying to take flight or lift into the air, winged, with all those white and silver feathers in motion. But I knew none of those species nested on the island; they only used it as a place for respite and feeding.

How would I find the solitary nest of an Oystercatcher? They don't raise their young in colonies as Least Terns and Brown Pelicans do, so their scrapes in the sand are much harder to spot.

The first clue was a Turkey Vulture skimming the sand, tilting on the heat of the barest updraft. Right on its tail, three American Oystercatchers piped loudly and physically pushed the vulture out of the island's airspace. I'd never seen Oystercatchers on the offense before, so I guessed there were chicks or eggs close by.

I crouched low and crept up the mound in the center of the island. Through the white sand's shimmer, I spotted a pair of birds, but not before they saw me. How fearsome I was to those parents with my spider-like tripod and upright slow-moving form. Imagine if the only way we could protect our newly born was to draw the predator away with our own bodies and our own voices, implying *there's no nest, no chicks, no eggs, keep your eye on me, let me draw you far away from what I am trying to bring into the world.*

They scrambled to their feet and I knew that before I could settle my eye on the nest site, the birds would distract me away with quick wing beats. Before I could guess where their treasure lay, the movement of their bodies would draw my eye to the left, tricking me from the nest. I withdrew, marking in my mind a post I might use to locate it with my optics. After I'd backed around the mound to a distance more acceptable to the parents, I set up my scope: Yes! Three eggs in the sand.

I continued my survey and found a second Oystercatcher brooding her eggs on the sand. Over her long orange dagger of a bill, through scarlet-rimmed eyes, she had been tracking my approach. Her eye saw my paddle slicing the quiet waters of the boat ramp, watched my path unfolding, while I myself still ascertained the mood of the wind. Never should I think that my eyes are brighter and more alert than hers, she who has sifted into this landscape every day of her life, and every day of the lives of her kind, for millions of years. From that long perspective, she watched me from her nest scrape on the sand, three eggs burning into her belly through her brood patch. Her job was to watch for danger.

Was I danger, was I really?

Need her pulse sharpen, need she spring off the nest, exposing her eggs to the sun, in order to draw my eye off the little shingle of beach that was home for her shell-bound brood?

Correct. Yes. Absolutely.

Yes. Our human selves are a grave danger to everything wild and vulnerable on the planet.

On a berm at the southern edge of the island I noted a third pair of extremely wary adult Oystercatchers, and my eye began to perceive the birds' defensive ploys. They had slipped down across the broad spoil's shoulder to watch me and wait out my intentions.

Their nest cradled a single mottled egg. I squatted on the sand to wonder over it but I didn't allow myself to touch the egg. No one was watching (no person, I mean), and I could have. Still, I curbed my reach, never extended my arm. I think my desire to touch was connected to my human evolution, back to a time when eggs answered the need for protein, and the practice of hunting birds' nests was solely linked to food.

I felt the anxiety of the pair who tended this nest, standing up the hill. If I were willing to invoke restraint, perhaps I could be almost an equal sensory partner. Our roles were so very different: I was the one who watched and wanted to know, and they were the objects I studied and counted and adored. Perhaps a relationship could be created if I agreed to curb my desire to be close, to back away, and honor their subjectivity. It could be born if I acknowledged their agency, and the fact that they were engaged in the serious business of continuing their kind on the planet. I intuited the moment when I had nearly exhausted them with my insistence on being in their space. I felt their signal: *Go now!* they said.

And so away I went around the island, thinking about those three nests spaced the length of a football field from one another, in an equilateral triangle as far apart as the dry land could serve.

My kayak scudded back to the ramp on the rising tide. The wind helped too, shoving into a line of black clouds building upriver. Glancing back, I watched a lone Oystercatcher, just an ember on the beach, keeper of the third nest I'd found. I knew the waypoints to return to her home: a single white post; an enormous log wedged into the sand; and a spread of white morning glories, blooming and perhaps scenting her long incubation with the company of their lovely orbs.

Weeks passed. On a mid-May morning, I was lucky to catch a slack tide on my weekly survey trip to the spoil island, which made for an easy paddle across the river's wide current. Nevertheless, sweat rolled down my neck and back, pasting my shirt to my skin. At either tip of the island, I watched Brown Pelicans and terns loafing, knee deep in the shallows.

Last week, I'd hit the tern jackpot. The high mound of soil in the center of the island was dotted with nesting Least Terns, and a few gull-billed. I'd assembled my spotting scope quickly, so that I could make a rough count and leave them alone, for they were panicked, and let me know of their displeasure by circling and calling and dive-bombing my head. At least thirty birds protected nests. As soon as I'd tallied them, and clicked off a few pictures on my phone to document eggs, I'd squirmed off the spoil hill and walked back to my boat. I'd been thinking about them ever since, how they scrapped with one another over nesting space, delivered small fish to their mates, kept a group watch for predators.

This day, I'd saved the center mound, the heart of the nesting tern territory, for last. When you survey nesting shorebirds, you watch for a spatial pattern. Least Terns bound and rebound from the sand, that pattern is what cues you to locate the black heads of parents tending their eggs on the sand. The week before, I had counted thirty nesting pairs, along with several Black Skimmers and some Gull-billed Terns, and of course, the three Oystercatcher nests around the edge. All week long, I'd been revisiting them in my mind.

Never did I imagine the terns would be gone. But it was true. The interior of the island, where spoil had been piled highest, was swept clean of birds. Briefly, I wondered if I had really seen them the week before. But I had a picture of eggs on nest to prove it, and bird poop on my hat, besides. And there had been many, many pairs. Adjusting my binoculars, I studied the ground, pausing to look closely at bits of plastic and other detritus shining in the sun, on the sand. Could I have mistaken that garbage for terns on the ground? No way. The Least Terns had been everywhere, dive-bombing, chattering, and warning me away. I scoured the hill with my scope, west to east.

But there wasn't a single tern on the mound. What happened to those birds? I chastised myself: I lived too far away to guard them more closely.

Until I had reached my mid-twenties, I didn't know the Earth was afflicted by our species. All I wanted was to submerge myself in the delight of it, and I did: the Atlantic Ocean, cold and dark and irresistible. Piles of autumn leaves: scarlet, orange, cadmium yellow. Canoe expeditions through the pitcher plant bogs of the Okefenokee Swamp, and the chill of the Suwannee River's springs. I took those wild places, and the reliable turn of the seasons, for granted. Excesses of winter simply meant a pair of snow days, summer, a brief wave of heat. There was no reason to imagine the seasons would ever lose the structure they offered my life. The natural world was mine to dwell upon, and I did.

But now I understand that all of the ways the coast and its islands, the springs and the native birds, the panthers and the Everglades, are being diminished, are symptoms of our culture's commitment to infinite growth. Infinite taking isn't possible on a finite planet. Our actions are both unethical and rife with repercussions, not just for future generations of humans, but for all species with whom we share the Earth. Including Oystercatchers.

I raged back across the river, my heart beating wildly. Back in my truck, I dialed up a local biologist in her office, from my cell.

"Such a huge bummer, Megan. There wasn't a single Least Tern on the island just now," I said. "You know I saw thirty birds on the ground incubating just a week ago, but today there are none at all."

I could hear her intake of breath. She was disappointed, too.

"Why, do you think?" I rushed along with my own train of thought. "Was that rain last Thursday enough to flood the nests, even up high on the spoil? Or, could it have been predators? Raccoon? What do you think?"

"Oh, Sue, I'm afraid I can guess what happened," Megan replied. "My worst fear. Last weekend, someone organized an event called Paddlejam, right there in Apalachicola. The idea was to try and break the world record of number of kayaks rafted together. It was billed as a fundraiser for at-risk youth by the local Methodist churches."

She continued. "I received several frantic phone calls on Saturday saying that the kayakers were told to stage on our spoil island before they made their kayak raft. By the time I got there, the kayak raft had already happened and no one was on island. I hoped the crowd hadn't been as bad as I imagined. Since the birds abandoned their nests, I'm afraid it was."

My heart sank; I dropped my forehead to the steering wheel of the truck, so hot it felt like a brand. It was a brand, the mark of the insufficient advocate.

And yet I knew that just like me when I was young, those paddlers had only good intentions, certainly harbored no desire to disrupt the nesting shorebirds on the spoil island. But there are just too many of us humans, and we don't (and don't know how to) calculate the needs of other species when we make our plans.

Seabirds and shorebirds nest alone or in colonies, depending on their species, in shallow scrapes. Sandy open beaches, between the high tide line and dune grasses are their homes, the only places they can nest and continue their kind. How many, many times I have watched beachgoers walk right past protective barriers, right past the most beautiful interpretive signage we can think to create, straight into the tiny spaces we have managed to set aside for the wild birds to nest. Not everyone can be told what to do and then simply left to their own devices. A commitment to kindness and respect has to come from some deeper cultural training; perhaps you have to be raised up in it. Somehow we must make this happen, because, as Kathleen Dean Moore has written, "This is the wonder-filled world that we are destroying, the lyric voices that we are silencing, the sanctity that we are defiling, at a rate and with a violence that cannot be measured."[1]

Until people change their minds, deeply, deeply change, and understand and respect the equivalent needs of every other species for life, all we can rely on is law-enforcement. The enforcers—wildlife officers and volunteer stewards (and there are never enough of them)—are all we have to keep people out of the sacred nesting places. Otherwise, ignorance and the desire to do whatever one pleases rules, and the birds suffer.

"My heart sinks when the Least Terns arrive in April and begin breeding—or trying to," a shorebird biologist told me recently. I understand his trepidation: the little birds' pugnacity is no match for beach driving trucks, dogs, fireworks, and every manner of human intrusion into their nesting beaches. Shorebirds and seabirds are declining everywhere they exist, and some are already gone.

Like so many wild birds, Least Tern populations were destroyed by hunters who shot them, to use their feathers to adorn women's hats. When the Migratory Bird Treaty Act was passed in 1916 and people began to change their attitudes toward conservation, Least Terns bounced back. But now, they are again so diminished by recreational, industrial, and residential development in their coastal breeding areas, they are specially classified for protection in much of their North American range. No other wide-ranging tern on our continent can claim that unfortunate distinction.

Near the end of the nesting season, I made my way around the tiny landscape with my spotting scope over my shoulder and my binoculars around my neck, as usual, keeping count of all the birds I saw. I advanced carefully over the sand to a place triangulated between a broken bit of plastic bucket and a certain white morning glory blossom.

And there they were: three eggs marbled brown and black, as fragile and unlikely as snowflakes on the sand. An extra high tide could so easily wash them away. A crow or a large gull could devour them. Or if all went well, this line of Oystercatchers might continue another generation.

In the last of the three nests, I witnessed life crossing between the worlds.

It was the tip of the bill of an Oystercatcher chick, meeting the salt air for the very first time. At first, I thought the hole meant that the egg must be damaged. Had ants punctured it, or was the eggshell thinned and then fractured by the weight of the parents' bodies? But no: there at the center of the hole, a tiny bill, a new rare life, was entering our world.

I allowed myself only the barest moment to ascertain what I saw, for the adults were circling wide, silent arcs out over the water and back. I wasn't welcome to watch, but I knew the full hatch of that chick, its first tumble onto the sand, would be as miraculous and sacred as the birth of any other creature on earth.

I needed to leave the eggs under the protection of their parents, even though my instinct was to kneel down and stay. Kneel down and pray.

As the sun began to angle into the sea, I returned to my kayak at the island's edge, thinking about how our planet and our sun had created such palettes uncounted nightfalls, long before the shorebirds or I were born. Earth has turned in far lonelier eons, without bird or human. I staggered under gratitude so weighty, I had to sink to my heels. The flaming sun lit up the whole of my face. The wind lifted my hair. I closed my eyes and became simply another breathing presence on the sand. No boundaries.

In our saltwater and bones bodies, each one of us loves the birds this much. They have companioned us through the Holocene, weavers of air current and nest cup. Their songs were our first music as a species, their call notes the first living patterns on our collective human eardrum. We learned percussion from the woodpecker, and to scream from the eagle, and to sing complicated melody from the warbler and the thrush.

Just as our bodies are constructed of the dust of stars, they also carry the memory of a time when our lives were always among the birds, under the spread of the sky. A time when we lived without separation. Would we humans knowingly let wild birds fall extinct? Not if we understood the repercussions of our actions, I don't think we would. Our Western legal and economic systems reinforce that the natural world is ours to exploit, not an ecological partner with its own rights to live and thrive. We must address the deep pathology of our time, the belief that our rights as human beings are different and carry more weight than those of the rest of creation. As the birds fare, so do we, and so do our grandchildren, and all the generations waiting for their chance at life.

Chapter 26

My Bird Problem

Jonathan Franzen

February 2005, South Texas: I'd checked into a roadside motel in Brownsville and was getting up in the dark every morning, making coffee for my old friend Manley, who wouldn't talk to me or leave his bed until he'd had some, and then bolting the motel's free breakfast and running to our rental car and birding nonstop for twelve hours. I waited until nightfall to buy lunch food and fill the car with gas, to avoid wasting even a minute of birdable daylight. The only way not to question what I was doing, and why I was doing it, was to do absolutely nothing else.

At the Santa Ana National Wildlife Refuge, on a hot weekday afternoon, Manley and I hiked several miles down dusty trails to an artificial water feature on the far margin of which I saw three pale-brown ducks. Two of them were paddling with all deliberate speed into the cover of dense reeds, affording me a view mainly of their butts, but the third bird loitered long enough for me to train my binoculars on its head, which looked as if a person had dipped two fingers in black ink and drawn horizontal lines across its face.

"A masked duck!" I said. "You see it?"

"I see the duck," Manley said.

"A masked duck!"

The bird quickly disappeared into the reeds and gave no sign of reemerging. I showed Manley its picture in my *Sibley*.

"I'm not familiar with this duck," he said. "But the bird in this picture is the one I just saw."

"The stripes on its face. The sort of cinnamony brown."

"Yes."

"It was a masked duck!"

We were within a few hundred yards of the Rio Grande. On the other side of the river, if you traveled south—say, to Brazil—you could see masked ducks by the dozens. They were a rarity north of the border, though. The pleasure of the sighting sweetened our long tramp back to the parking lot.

While Manley lay down in the car to take a nap, I poked around in a nearby marsh. Three middle-aged white guys with good equipment asked me if I'd seen anything interesting.

"Not much," I said, "except a masked duck."

All three began to talk at once.

"A masked duck!"

"Masked duck?"

"Where exactly? Show us on the map."

"Are you sure it was a masked duck?"

"You're familiar with the ruddy duck. You do know what a female ruddy looks like."

"Masked duck!"

I said that, yes, I'd seen female ruddies, we had them in Central Park, and this wasn't a ruddy duck. I said it was as if somebody had dipped two fingers in black ink and—

"Was it alone?"

"Were there others?"

"A masked duck!"

One of the men took out a pen, wrote down my name, and had me pinpoint the location on a map. The other two were already moving down the trail I'd come up.

"And you're sure it was a masked duck," the third man said.

"It wasn't a ruddy," I said.

A fourth man stepped out of some bushes right behind us. "I've got a nighthawk sleeping in a tree."

"This guy saw a masked duck," the third man said.

"A masked duck! Are you sure? Are you familiar with the female ruddy?"

The other two men came hurrying back up the trail. "Did someone say nighthawk?"

"Yeah, I've got a scope on it."

The five of us went into the bushes. The nighthawk, asleep on a tree branch, looked like a partly balled gray hiking sock. The scope's owner said that the friend of his who'd first spotted the bird had called it a lesser nighthawk, not a common. The well-equipped trio begged to differ.

"He said lesser? Did he hear its call?"

"No," the man said. "But the range—"

"Range doesn't help you."

"Range argues for common, if anything, at this time of year."

"Look where the wing bar is."

"Common."

"Definitely calling it a common."

The four men set off at a forced-march pace to look for the masked duck, and I began to worry. My identification of the duck, which had felt ironclad in the moment, seemed dangerously hasty in the context of four serious birders marching several miles in the afternoon heat. I went and woke up Manley.

"The only thing that matters," he said, "is that we saw it."

"But the guy took my name down. Now, if they don't see it, I'm going to get a bad rep."

"If they don't see it, they'll think it's in the reeds."

"But what if they see ruddies instead? There could be ruddies *and* masked ducks, and the ruddies aren't as shy."

"It's something to be anxious about," Manley said, "if you want to be anxious about something."

I went to the refuge visitor center and wrote in the logbook: *One certain and two partially glimpsed MASKED DUCKS, north end of Cattail* #2. I asked a volunteer if anyone else had reported a masked duck.

"No, that would be our first this winter," she said.

The next afternoon, on South Padre Island, in the wetland behind the Convention Center, where about twenty upper-Midwestern retirees and scraggly-bearded white guys were pacing the boardwalks with cameras and binoculars, I saw a pretty, dark-haired young woman taking telephoto pictures of a pair of ducks. "Green-winged teals," I mentioned to Manley.

The girl looked up sharply. "Green-winged teals? Where?"

I nodded at her birds.

"Those are wigeons," she said.

"Right."

I'd made this mistake before. I knew perfectly well what a wigeon looked like, but sometimes, in the giddiness of spotting something, my brain got confused. As Manley and I retreated down the boardwalk, I showed him pictures.

"See," I said, "the wigeon and the green-winged teal have more or less the same palette, just completely rearranged. I should have said wigeon. Now she thinks I can't tell a wigeon from a teal."

"Why didn't you just tell her that?" Manley said. "Just say that the wrong word came out of your mouth."

"That would only have compounded it. It would have been protesting too much."

"But at least she'd know you know the difference."

"She doesn't know my name. I'll never see her again. That is my only conceivable consolation."

There is no better American place for birds in February than South Texas. Although Manley had been down here thirty years earlier, as a teenage birder, it was a wholly new world to me. In three days, I'd seen fetchingly disheveled anis flopping around on top of shrubs, Jurassic-looking anhingas sun-drying their wings, squadrons of white pelicans gliding downriver on nine-foot wingspans, a couple of caracaras eating a road-killed king snake, an elegant trogon and a crimson-collared grosbeak and two exotic robins all lurking on a postage-stamp Audubon Society tract in Weslaco. The only frustration was my No. 1 trip target bird, the black-bellied whistling-duck. A tree nester, strangely long-legged, with a candy-pink bill and a bold white eye ring, the whistling-duck was one of those birds in the field guide which I couldn't quite believe existed—something out of Marco Polo. It was supposed to winter in good numbers on Brownsville's urban oxbow lakes (called *resacas*), and with each shoreline that I scanned in vain, the bird became that much more mythical to me.

Out on South Padre, as fog rolled in off the Gulf of Mexico, I remembered to look up at the city water tower, where, according to my guidebook, a peregrine falcon often perched. Sure enough, very vaguely, I saw the peregrine up there. I set up my spotting scope, and an older couple, two seasoned-looking birders, asked me what I had.

"Peregrine falcon," I said proudly.

"You know, Jon," Manley said, his eye to the scope, "the head looks more like an osprey."

"That is an osprey," the woman quietly affirmed.

"*God*," I said, looking again, "it is *so hard* to tell in the fog, and to get a sense of scale, you know, way up there, but you're right, yes, I see it. Osprey, osprey, osprey. Yes."

"That's the great thing about fog," the woman remarked. "You can see whatever you want."

Just then the dark-haired young woman came by with her tripod and big camera.

"Osprey," I told her confidently. "By the way, you know, I'm still totally writhing about saying 'teal' when I meant 'gadwall.'"

She stared at me. "*Gadwall?*"

Back in the car, using Manley's phone to avoid betraying my own name via caller ID, I called the visitor center at Santa Ana and asked if "people" had been reporting any masked ducks on the refuge.

"Yes, somebody did report one yesterday. Down at Cattails."

"Just one person?" I asked.

"Yes. I wasn't here. But somebody did report a masked duck."

"Fantastic!" I said—as if, by sounding excited, I could lend after-the-fact credibility to my own report. "I'll come look for it!"

Halfway back to Brownsville, on one of the narrow dirt roads that Manley liked to direct me down, we stopped to admire a lushly green-girdled blue *resaca* with the setting sun behind us. The delta in winter was too beautiful to stay embarrassed in for long. I got out of the car, and there, silent, on the shadowed side of the water, floating nonchalantly, as if it were the most natural thing in the world—which is, after all, the way of magical creatures in enchanted places—was my black-bellied whistling-duck.

It felt weird to return to New York. After the excitements of South Texas, I was hollow and restless, like an addict in withdrawal. It was a chore to make myself comprehensible to friends; I couldn't keep my mind on my work. Every night, I lay down with bird books and read about other trips I could take, studied the field markings of species I hadn't seen, and then dreamed vividly of birds. When two kestrels, a male and a female, possibly driven out of Central Park by the artist Christo and his wife, Jeanne-Claude, began showing up on a chimney outside my kitchen window and bloodying their beaks on fresh-killed mice, their dislocation seemed to mirror my own.

One night in early March, I went to the Society for Ethical Culture to hear Al Gore speak on the subject of global warming. I was expecting to be amused by the speech's rhetorical badness—to roll my eyes at Gore's intoning of "fate" and "mankind," his flaunting of his wonk credentials, his scolding of American consumers. But Gore seemed to have rediscovered a sense of humor. His speech was fun to listen to, if unbelievably depressing. For more than an hour, with heavy graphical support, he presented compelling evidence of impending climate-driven cataclysms that will result in unimaginable amounts of upheaval and suffering around the globe, possibly within my own lifetime. I left the auditorium under a cloud of grief and worry of the sort I'd felt as a teenager reading about nuclear war.

Ordinarily, in New York, I keep a tight rein on my environmental consciousness, confining it, ideally, to the ten minutes per year when I write my guilt-assuaging checks to groups like the Sierra Club. But Gore's message was so disturbing that I was nearly back to my apartment before I could think of some reasons to discount it. Like: wasn't I already doing more than most Americans to combat global warming? I didn't own a car, I lived in an energy-efficient Manhattan apartment, I was good about recycling. Also: wasn't the weather that night *unusually cold* for early March? And hadn't Gore's maps of Manhattan in the future, the island half-submerged by rising sea levels, all shown that the corner of Lexington and Eighty-first Street, where I live, would stay high and dry in even the worst-case scenario? The Upper East Side has a definite topography. It seemed unlikely that seawater from Greenland's melting ice cap would advance any farther than the Citarella market on Third Avenue, six blocks to the south and east. Plus, my apartment was way up on the tenth floor.

When I went inside, no kids came running to meet me, and this absence of kids seemed to clinch it: I was better off spending my anxiety budget on viral pandemics and dirty bombs than on global warming. Even if I had had kids, it would have been hard work for me to care about the climatic well-being of their children's children. Not having kids freed me altogether. Not having kids was my last, best line of defense against the likes of Al Gore.

There was only one problem. Trying to fall asleep that night, mentally replaying Gore's computer images of a desertified North America, I couldn't find a way not to care about the billions of birds and thousands of avian species that were liable to be wiped out worldwide. Many of the Texan places that I'd visited in February had elevations of less than twenty feet, and the climate down there was already almost lethally extreme. Human beings could probably adapt to future changes, we were famously creative at averting disasters and at making up great stories when we couldn't, but birds didn't have our variety of options. Birds needed help. And this, I realized, was the true disaster for a comfortable modern American. This was the scenario I'd been at pains to avert for many years: not the world's falling apart in the future, but my feeling inconveniently obliged to care about it in the present. This was my bird problem.

For a long time, back in the eighties, my wife and I lived on our own little planet. We spent thrilling, superhuman amounts of time by ourselves. In our first two apartments, in Boston, we were so absorbed in each other that we got along with exactly one good friend, our college classmate Ekström, and when we finally moved to Queens, Ekström moved to Manhattan, thereby sparing us the need to find a different friend.

Early in our marriage, when my old German instructor Weber asked me what the two of us were doing for a social life, I said we didn't have one. "That's sweet for a year," Weber said. "Two years at the most." His certainty offended me. It struck me as extremely condescending, and I never spoke to him again.

None of the doom criers among our relatives and former friends, none of these brow-furrowing emotional climatologists, seemed to recognize the special resourcefulness of our union. To prove them wrong, we made our aloneness work for four years, for five years, for six years; and then, when the domestic atmosphere really did begin to overheat, we fled from New York to a Spanish village where we didn't know anybody and the villagers hardly even spoke Spanish. We were like those habit-bound peoples in Jared Diamond's *Collapse* who respond to an ecosystem's degradation by redoubling their demands on it—medieval Greenlanders, prehistoric Easter Islanders, contemporary SUV buyers. Whatever reserves the two of us still had when we arrived in Spain were burned up in seven months of isolation.

Returning to Queens, we could no longer stand to be together for more than a few weeks, couldn't stand to see each other so unhappy, without running somewhere else. We reacted to minor fights at breakfast by lying face down on the floor of our respective rooms for hours at a time, waiting for acknowledgment of our pain. I wrote poisonous jeremiads to family members who I felt had slighted my wife; she presented me with handwritten fifteen- and twenty-page analyses of our condition; I was putting away a bottle of Maalox every week. It was clear to me that something was terribly wrong. And what was wrong, I decided, was modern industrialized society's assault on the environment.

In the early years, I'd been too poor to care about the environment. My first car in Massachusetts was a vinyl-top '72 Nova that needed a tailwind to achieve ten miles a gallon and whose exhaust was boeuf bourguignon—like in its richness and complexity. After the Nova died, we got a Malibu wagon whose ridiculous four-barrel carburetor ($800) needed replacing and whose catalytic converter ($350) had had its guts scraped out to ease the flow of gases. Polluting the air a little less would have cost us two or three months' living expenses. The Malibu practically knew its own way to the crooked garage where we bought our annual smog-inspection sticker.

The summer of 1988, however, had been one of the hottest on record in North America, and rural Spain had been a spectacle of unchecked development and garbage-strewn hillsides and diesel exhaust, and after the dismantlement of the Berlin Wall the prospect of nuclear annihilation (my longtime pet apocalypse) was receding somewhat, and the great thing about the rape of nature, as an alternative apocalypse, was the opportunity it gave me to blame myself. I had grown up listening to daily lectures on personal responsibility. My father was a saver of string and pencil stubs and a bequeather of fantastic Swedish Protestant prejudices. (He considered it unfair to drink a cocktail at home before going to a restaurant, because restaurants depended on liquor sales for profits.) To worry about the Kleenexes and paper towels I was wasting and the water I was letting run while I shaved and the sections of the Sunday *Times* I was throwing away unread and the pollutants I was helping to fill the sky with every time I took an airplane came naturally to me.

I argued passionately with a friend who believed that fewer BTUs were lost in keeping a house at 68 degrees overnight than in raising the temperature to 68 in the morning. Every time I washed out a peanut-butter jar, I tried to calculate whether less petroleum might be used in manufacturing a new jar than in heating the dishwater and transporting the old jar to a recycling center.

My wife moved out in December 1990. A friend had invited her to come and live in Colorado Springs, and she was ready to escape the pollution of her living space by me. Like modern industrialized society, I continued to bring certain crucial material benefits to our household, but these benefits came at an ever greater psychic cost. By fleeing to the land of open skies, my wife hoped to restore her independent nature, which years of too-married life had compromised almost beyond recognition. She rented a pretty apartment on North Cascade Avenue and sent me excited letters about the mountain weather. She became fascinated with narratives of pioneer women—tough, oppressed, resourceful wives who buried dead infants, watched freak June blizzards kill their crops and livestock, and survived to write about it. She talked about lowering her resting pulse rate below thirty.

Back in New York, I didn't believe we'd really separated. It may have become impossible for us to live together, but my wife's sort of intelligence still seemed to me the best sort, her moral and aesthetic judgments still seemed to me the only ones that counted. The smell of her skin and the smell of her hair were restorative, irreplaceable, the best. Deploring other people—their lack of perfection—had always been our sport. I couldn't imagine never smelling her again.

The next summer, we went car-camping in the West. I was frankly envious of my wife's new Western life, and I also wanted to immerse myself in nature, now that I'd become environmentally conscious. For a month, the two of us followed the retreating snow up through the Rockies and

Cascades, and made our way back south through the emptiest country we could find. Considering that we were back together 24/7, sharing a small tent, and isolated from all social contacts, we got along remarkably well.

What sickened and enraged me were all the other human beings on the planet. The fresh air, the smell of firs, the torrents of snowmelt, the columbines and lupine, the glimpses of slender-ankled moose were nice sensations, but not intrinsically any nicer than a gin martini or a well-aged steak. To really deliver the goods, the West also had to conform to my wish that it be unpopulated and pristine. Driving down an empty road through empty hills was a way of reconnecting with childhood fantasies of being a Special Adventurer—of feeling again like the children in Narnia, like the heroes of Middle-earth. But house-sized tree pullers weren't clear-cutting Narnia behind a scrim of beauty strips. Frodo Baggins and his compatriots never had to share campgrounds with forty-five identical Fellowships of the Ring wearing Gore-Tex parkas from REI. Every crest in the open road opened up new vistas of irrigation-intensive monoculture, mining-scarred hillsides, and parking lots full of nature lovers' cars. To escape the crowds, my wife and I took longer hikes in deeper backcountry, toiling through switchbacks, only to find ourselves on dusty logging roads littered with horse manure. And here—look out!—came some gonzo clown on his mountain bike. And there, overhead, went Delta Flight 922 to Cincinnati. And here came a dozen Boy Scouts with jangling water cups and refrigerator-sized backpacks. My wife had her cardiovascular ambitions to occupy her, but I was free to stew all day long: Were those human voices up ahead? Was that a speck of aluminum foil in the tree litter? Or, oh no, were those human voices coming up *behind us*?

I stayed in Colorado for a few more months, but being in the mountains had become unbearable to me. Why stick around to see the last beautiful wild places getting ruined, and to hate my own species, and to feel that I, too, in my small way, was one of the guilty ruiners? In the fall I moved back East. Eastern ecologies, specifically Philadelphia's, had the virtue of already being ruined. It eased my polluter's conscience to lie, so to speak, in a bed I'd helped to make. And this bed wasn't even so bad. For all the insults it had absorbed, the land in Pennsylvania was still riotously green.

The same could not be said of our marital planet. There, the time had come for me to take decisive action; the longer I delayed, the more damage I would do. Our once limitless-seeming supply of years for having kids, for example, had suddenly and alarmingly dwindled, and to dither for even just a few more years would be permanently ruinous. And yet: what decisive action to take? At this late date, I seemed to have only two choices. Either I should try to change myself radically—devote myself to making my wife happy, try to occupy less space, and be, if necessary, a full-time dad—or else I should divorce her.

Radically changing myself, however, was about as appetizing (and likely to happen) as volunteering for the drab, homespun, post-consumerist society that the "deep ecologists" tell us is the only long-term hope for humans on the planet. Although I talked the talk of fixing and healing, and sometimes I believed it, a self-interested part of me had long been rooting for trouble and waiting, with calm assurance, for the final calamity to engulf us. I had old journals containing transcripts of early fights which read word-for-word like the fights we were having ten years later. I had a carbon copy of a letter I'd written to my brother Tom in 1982, after I'd announced our engagement to my family and Tom had asked me why the two of us didn't just live together and see how things went; I'd replied that, in the Hegelian system, a subjective phenomenon (e.g., romantic love) did not become, properly speaking, "real" until it took its place in an objective structure, and that it was therefore important that the individual and the civic be synthesized in a ceremony of commitment. I had wedding pictures in which, before the ceremony of commitment, my wife looked beatific and I could be seen frowning and biting my lip and hugging myself tightly.

But giving up on the marriage was no less unthinkable. It was possible that we were unhappy because we were trapped in a bad relationship, but it was also possible that we were unhappy for other reasons, and that we should be patient and try to help each other. For every doubt documented in the fossil record, I could find an old letter or journal entry in which I talked about our marriage with happy certainty, as if we'd been together since the formation of the solar system, as if there had always been the two of us and always would be. The skinny, tuxedoed kid in our wedding pictures, once the ceremony was over, looked unmistakably smitten with his bride.

So more study was needed. The fossil record was ambiguous. The liberal scientific consensus was self-serving. Maybe, if we tried a new city, we could be happy? We traveled to check out San Francisco, Oakland,

Portland, Santa Fe, Seattle, Boulder, Chicago, Utica, Albany, Syracuse, and Kingston, New York, finding things to fault in each of them. My wife came back and joined me in Philadelphia, and I borrowed money at interest from my mother and rented a three-story, five-bedroom house that neither of us could stand to live in by the middle of 1993. I sublet a place for myself in Manhattan which I then, out of guilt, handed over to my wife. I returned to Philadelphia and rented yet a third space, this one suitable for both working and sleeping, so that my wife would have all five of the house's bedrooms at her disposal, should she need them, on her return to Philly. Our financial hemorrhaging in late 1993 looked a lot like the country's energy policy in 2005. Our determination to cling to unsustainable dreams was congruent with—maybe even identical to—our drive to bankrupt ourselves as rapidly as possible.

Around Christmastime, the money ran out altogether. We broke our leases and sold the furniture. I took the old car, she took the new laptop, I slept with other people. Unthinkable and horrible and ardently wished-for: our little planet was ruined.

A staple of my family's dinner-table conversation in the mid-seventies was the divorce and remarriage of my father's boss at the railroad, Mr. German. Nobody of my parents' generation in either of their extended families had ever been divorced, nor had any of their friends, and so the two of them steeled each other in their resolve not to know Mr. German's young second wife. They exhaustively pitied the first wife, "poor Glorianna," who had been so dependent on her husband that she'd never even learned to drive. They expressed relief and worry at the Germans' departure from their Saturday-night bridge club, since Mr. German was bad at bridge but Glorianna was now left without a social life. One night my father came home and said he'd almost lost his job that day at lunch. In the executive dining room, while Mr. German and his subordinates were discussing how to assess a person's character, my father had found himself remarking that he judged a man by how he played a bridge hand. I wasn't old enough to understand that he hadn't really almost lost his job for this, or that condemning Mr. German and pitying Glorianna were ways for my parents to talk about their own marriage, but I did understand that dumping your wife for a younger woman was the sort of despicable selfish thing that a chronic overbidder might do.

A related talk staple in those years was my father's hatred of the Environmental Protection Agency. The young agency had issued complicated rules about soil pollution and toxic runoff and riverbank erosion, and some of the rules seemed unreasonable to my father. What really enraged him, though, were the enforcers. Night after night he came home fuming about these "bureaucrats" and "academics," these high-handed "so-and-sos" who didn't bother to hide how morally and intellectually superior they felt to the corporations they were monitoring, and who didn't think they owed explanations, or even basic courtesy, to people like my father.

The odd thing was how closely my father's values resembled those of his enemies. The breakthrough environmental legislation of that era, including the Clean Air and Clean Water Acts and the Endangered Species Act, had attracted the support of President Nixon and both parties in Congress precisely because it made sense to old-fashioned Protestants, like my parents, who abhorred waste and made sacrifices for their kids' future and respected God's works and believed in taking responsibility for their messes. But the social ferment that gave rise to the first Earth Day, in 1970, unleashed a host of other energies—the incivility of the so-and-sos, the pleasurable self-realizations of Mr. German, the cult of individuality—that were inimical to the old religion and ultimately won out.

Certainly I, as a self-realizing individual in the nineties, was having trouble with my parents' logic of unselfishness. Deprive myself of an available pleasure *why?* Take shorter and colder showers *why?* Keep having anguished phone conversations with my estranged wife on the subject of our failure to have children *why?* Struggle to read Henry James's last three novels *why?* Stay mindful of the Amazon rain forest *why?* New York City, which I returned to for good in 1994, was becoming a very pleasant place to live again. The nearby Catskills and Adirondacks were better protected than the Rockies and Cascades. Central Park, under recultivation by deep-pocketed locals, was looking greener every spring, and the other people out walking in it didn't enrage me: this was a *city*; there were *supposed* to be other people. On a May night in 1996, I walked across the park's newly restored, deep-pile lawns to a party where I saw a beautiful and very young woman standing awkwardly in a corner, behind a floor lamp that she twice nearly knocked over, and I felt so liberated that

I could no longer remember one single reason not to introduce myself to her and, in due course, start asking her out.

The old religion was finished. Without its cultural support, the environmental movement's own cult of wilderness was never going to galvanize mass audiences. John Muir, writing from San Francisco at a time when you could travel to Yosemite without hardship and still have the valley to yourself for spiritual refreshment, founded a religion that required a large parcel of empty backcountry for every worshipper. Even in 1880, there weren't enough parcels like this to go around. Indeed, for the next eighty years, until Rachel Carson and David Brower sounded their populist alarms, the preserving of wild nature was generally assumed to be the province of elites. The organization that Muir formed to defend his beloved Sierras was a Club, not an Alliance. Henry David Thoreau, whose feelings for pine trees were romantic, if not downright sexual, called the workers who felled them "vermin." For Edward Abbey, who was the rare green writer with the courage of his misanthropy, the appeal of southeastern Utah was, frankly, that its desert was inhospitable to the great herd of Americans who were incapable of understanding and respecting the natural world. Bill McKibben, Harvard graduate, followed up his apocalyptic *The End of Nature* (in which he contrasted his own deep reverence for nature with the shallow-minded "hobby" that nature is for most outdoorspeople) with a book about cable TV's inferiority to the timeless pleasures of country living. To Verlyn Klinkenborg, the professional trivialist whose job is to remind *New York Times* readers that spring follows winter and summer follows spring, and who sincerely loves snowdrifts and baling twine, the rest of humanity is a distant blob notable for its "venality" and "ignorance."

And so, once the EPA had cleaned up the country's most glaring messes, once sea otters and peregrine falcons had rebounded from near extinction, once Americans had had a disagreeable taste of European-style regulation, the environmental movement began to look like just another special interest hiding in the skirts of the Democratic Party. It consisted of well-heeled nature enthusiasts, tree-spiking misanthropes, nerdy defenders of unfashionable values (thrift, foresight), invokers of politically unfungible abstractions (the welfare of our great-grandchildren), issuers of shrill warnings about invisible risks (global warming) and exaggerated hazards (asbestos in public buildings), tiresome scolds about consumerism, reliers on facts and policies in an age of image, a constituency loudly proud of its refusal to compromise with others.

Bill Clinton, the first boomer President, knew a stinker when he saw one. Unlike Richard Nixon, who had created the EPA, and unlike Jimmy Carter, who had set aside twenty-five million acres of Alaska as permanent wilderness, Clinton needed the Sierra Club a lot less than it needed him. In the Pacific Northwest, on lands belonging to the American people, the U.S. Forest Service was spending millions of tax dollars to build roads for multinational timber companies that were clear-cutting gorgeous primeval forests and taking handsome profits for themselves, preserving a handful of jobs for loggers who would soon be out of work anyway, and shipping much of the timber to Asia for processing and sale. You wouldn't think this issue was an automatic public-relations loser, but groups like the Sierra Club decided to fight the battle out of public sight, in federal court, where their victories tended to be Pyrrhic; and the boomer President, whose need for love was nonsatiable by Douglas firs or spotted owls but conceivably could be met by lumberjacks, soon added the decimation of the Northwest's old-growth forests to a long list of related setbacks—an environmentally toothless NAFTA, the metastasis of exurban sprawl, the lowering of average national vehicle fuel efficiency, the triumph of the SUV, the accelerating depletion of the world's fisheries, the Senate's 95-0 demurral on the Kyoto Protocol, etc.—in the decade when I left my wife and took up with a twenty-seven-year-old and really started having fun.

Then my mother died, and I went out birdwatching for the first time in my life. This was in the summer of 1999. I was on Hat Island, a wooded loaf of gravel subdivided for small weekend homes, near the blue-collar town of Everett, Washington. There were eagles and kingfishers and Bonaparte's gulls and dozens of identical sparrows that persisted, no matter how many times I studied them, in resembling six different sparrow species in the field guide I was using. Flocks of goldfinches brilliantly exploded up over the island's sunlit bluffs like something ceremonial and Japanese. I saw my first northern flicker and enjoyed its apparent confusion about what kind of bird it was. Unwoodpeckerish in plumage, like a mourning dove in war paint, it flew dippingly, in typical woodpecker fashion, white rump flashing, from one ill-fitting identity to another. It had a way of landing with a little crash wherever. In its careening beauty it reminded me of my former girlfriend, the one I'd first glimpsed tangling with a floor lamp and was still very fond of, though from a safe remove now.

I had since met a vegetarian Californian writer, a self-described "fool for animals," slightly older than I, who had no discernible interest in getting pregnant or married or in moving to New York. As soon as I'd fallen for her, I'd set about trying to change her personality and make it more like mine; and although, a year later, I had nothing to show for this effort, I at least didn't have to worry about ruining somebody again. The Californian was a veteran of a ruinous marriage of her own. Her indifference to the idea of kids spared me from checking my watch every five minutes to see if it was time for my decision about her reproductive future. The person who wanted kids was me. And, being a man, I could afford to take my time.

The last day I ever spent with my mother, at my brother's house in Seattle, she asked me the same questions over and over: Was I pretty sure that the Californian was the woman I would end up with? Did I think we would probably get married? Was the Californian actually divorced yet? Was she interested in having a baby? Was I? My mother was hoping for a glimpse of how my life might proceed after she was gone. She'd met the Californian only once, at a noisy restaurant in Los Angeles, but she wanted to feel that our story would continue and that she'd participated in it in some small way, if only by expressing her opinion that the Californian really ought to be divorced by now. My mother loved to be a part of things, and having strong opinions was a way of not feeling left out. At any given moment in the last twenty years of her life, family members in three time zones could be found worrying about her strong opinions or loudly declaring that they didn't care about them or phoning each other for advice on how to cope with them.

Whoever imagined that LOVE YOUR MOTHER would make a good environmental bumper sticker obviously didn't have a mom like mine. Well into the nineties, tailing Subarus or Volvos outfitted with this admonition and its accompanying snapshot of Earth, I felt obscurely hectored by it, as if the message were "Nature Wonders Why She Hasn't Heard from You in Nearly a Month" or "Our Planet *Strongly* Disapproves of Your Lifestyle" or "The Earth Hates to Nag, But . . ." Like the natural world, my mother had not been in the best of health by the time I was born. She was thirty-eight, she'd had three successive miscarriages, and she'd been suffering from ulcerative colitis for a decade. She kept me out of nursery school because she didn't want to let go of me for even a few hours a week. She sobbed frighteningly when my brothers went off to college. Once they were gone, I faced nine years of being the last handy object of her maternal longings and frustrations and criticisms, and so I allied myself with my father, who was embarrassed by her emotion. I began by rolling my eyes at everything she said. Over the next twenty-five years, as she went on to have acute phlebitis, a pulmonary embolism, two knee replacements, a broken femur, three miscellaneous orthopedic surgeries, Raynaud's disease, arthritis, biannual colonoscopies, monthly blood-clot tests, extreme steroidal facial swelling, congestive heart failure, and glaucoma, I often felt terribly sorry for her, and I tried to say the right things and be a dutiful son, but it wasn't until she got a bad cancer diagnosis, in 1996, that I began to do what those bumper stickers admonished me to do.

She died in Seattle on a Friday morning. The Californian, who had been due to arrive that evening and spend some days getting acquainted with her, ended up alone with me for a week at my brother's vacation house on Hat Island. I broke down in tears every few hours, which I took as a sign that I was working through my grief and would soon be over it. I sat on the lawn with binoculars and watched a spotted towhee scratch vigorously in the underbrush, like somebody who really enjoyed yard work. I was pleased to see chestnut-backed chickadees hopping around in conifers, since, according to the guidebook, conifers were their favored habitat. I kept a list of the species I'd seen.

By midweek, though, I'd found a more compelling pastime: I began to badger the Californian about having children and the fact that she wasn't actually divorced yet. In the style of my mother, who had been a gifted abrader of the sensitivities of people she was unhappy with, I gathered and collated all the faults and weaknesses that the Californian had ever privately confessed to me, and I showed her how these interrelated faults and weaknesses were preventing her from deciding, *right now*, whether we would probably get married and whether she wanted to have children. By the end of the week, fully seven days after my mother's death, I was sure I was over the worst of my grief, and so I was mystified and angered by the Californian's unwillingness to move to New York and immediately try to get pregnant. Even more mystified and angered a month later, when she took wing to Santa Cruz and refused to fly back.

On my first visit to the cabin where she lived, in the Santa Cruz Mountains, I'd stood and watched mallards swimming in the San Lorenzo River. I was struck by how frequently a male and a female paired up, one waiting on the other while it nosed in the weeds. I had no intention of living without steak or bacon, but after that trip, as a token of vegetarianism, I decided to stop eating duck. I asked my friends what they knew about ducks. All agreed that they were beautiful animals; several also commented that they did not make good pets.

In New York, while the Californian took refuge from me in her cabin, I seethed with strong opinions. *The only thing I wanted* was for her and me to be in the same place, and I would gladly have gone out to California *if only she'd told me up front* that she wasn't coming back to New York. The more months that went by without our getting closer to a pregnancy, the more aggressively I argued for living together, and the more aggressively I argued, the flightier the Californian became, until I felt I had no choice but to issue an ultimatum, which resulted in a breakup, and then a more final ultimatum, which resulted in a more final breakup, and then a final final ultimatum, which resulted in a final final breakup, shortly after which I went out walking along the lake in lower Central Park and saw a male and a female mallard swimming side by side, nosing in the weeds together, and burst into tears.

It wasn't until a year or more later, after the Californian had changed her mind and come to New York, that I faced medical facts and admitted to myself that we weren't just going to up and have a baby. And even then I thought: Our domestic life is good right now, but if I ever feel like trying a different life with somebody else, I'll have a ready-made escape route from my current one: "Didn't I always say I wanted children?" Only after I turned forty-four, which was my father's age when I was born, did I get around to wondering why, if I was so keen to have kids, I'd chosen to pursue a woman whose indifference to the prospect had been clear from the beginning. Was it possible that I only wanted kids with this one particular person, because I loved her? It was apparent, in any case, that my wish for kids had become nontransferable. I was not Henry the Eighth. It wasn't as if I found fertility a lovable personality trait or a promising foundation for a lifetime of great conversation. On the contrary, I seemed to meet a lot of very boring fertile people.

Finally, sadly, around Christmastime, I came to the conclusion that my ready-made escape route had disappeared. I might find some other route later, but this route was no more. For a while, in the Californian's cabin, I was able to take seasonal comfort in stupefying amounts of aquavit, champagne, and vodka. But then it was New Year's, and I faced the question of what to do with myself for the next thirty childless years; and the next morning I got up early and went looking for the Eurasian wigeon that had been reported in south Santa Cruz County.

My affair with birds had begun innocently—an encounter on Hat Island, a morning of sharing binoculars with friends on Cape Cod. I wasn't properly introduced until a warm spring Saturday when the Californian's sister and brother-in-law, two serious birders who were visiting New York for spring migration, took me walking in Central Park. We started at Belvedere Castle, and right there, on mulchy ground behind the weather station, we saw a bird shaped like a robin but light-breasted and feathered in russet tones. A veery, the brother-in-law said.

I'd never even heard of veeries. The only birds I'd noticed on my hundreds of walks in the park were pigeons and mallards and, from a distance, beyond a battery of telescopes, the nesting red-tailed hawks that had become such overexposed celebrities. It was weird to see a foreign, unfamous veery hopping around in plain sight, five feet away from a busy footpath, on a day when half of Manhattan was sunning in the park. I felt as if, all my life, I'd been mistaken about something important. I followed my visitors into the Ramble in agreeably engrossed disbelief, as in a dream in which yellowthroats and redstarts and black-throated blue and black-throated green warblers had been placed like ornaments in urban foliage, and a film production unit had left behind tanagers and buntings like rolls of gaffer's tape, and ovenbirds were jogging down the Ramble's eroded hillsides like tiny costumed stragglers from some Fifth Avenue parade: as if these birds were just momentary bright litter, and the park would soon be cleaned up and made recognizable again.

Which it was. By June, the migration was over; songbirds were no longer flying all night and arriving in New York at dawn, seeing bleak expanses of pavement and window, and heading to the park for refreshment. But that Saturday afternoon had taught me to pay more attention. I started budgeting extra minutes when I had to cross the park to get somewhere. Out in the country, from the windows of generic motels, I looked at the cattails and sumac by interstate overpasses and wished I'd brought binoculars. A glimpse of dense brush or a rocky shoreline gave me an infatuated feeling, a sense of the world's being full of possibility. There were new birds to look for everywhere, and little by little I figured out the best hours (morning) and the best places (near some water) to go looking. Even then, it sometimes happened that I would walk through the park and see no bird more unusual than a starling, literally not one, and I would feel unloved and abandoned and wronged. (The stupid birds: where were they?) But then, later in the week, I'd see a spotted sandpiper by the Turtle Pond, or a hooded merganser on the Reservoir, or a green heron in some dirt by the Bow Bridge, and be happy.

Birds were what became of dinosaurs. Those mountains of flesh whose petrified bones were on display at the Museum of Natural History had done some brilliant retooling over the ages and could now be found living in the form of orioles in the sycamores across the street. As solutions to the problem of earthly existence, the dinosaurs had been pretty great, but blue-headed vireos and yellow warblers and white-throated sparrows—feather-light, hollow-boned, full of song—were even greater. Birds were like dinosaurs' better selves. They had short lives and long summers. We all should be so lucky as to leave behind such heirs.

The more I looked at birds, the more I regretted not making their acquaintance sooner. It seemed to me a sadness and a waste that I'd spent so many months out West, camping and hiking amid ptarmigans and solitaires and other fantastic birds, and had managed, in all that time, to notice and remember only one: a long-billed curlew in Montana. How different my marriage might have been if I'd been able to go birding! How much more tolerable our year in Spain might have been made by European waterfowl!

And how odd, come to think of it, that I'd grown up unscathed by Phoebe Snetsinger, the mom of one of my Webster Groves classmates, who later became the most successful birder in the world. After she was diagnosed with metastatic malignant melanoma, in 1981, Snetsinger decided to devote the remaining months of her life to really serious birding, and over the next two decades, through repeated remissions and recurrences, she saw more species than any other human being before or since; her list was near eighty-five hundred when she was killed in a road accident while chasing rarities in Madagascar. Back in the seventies, my friend Manley had come under Snetsinger's influence. He finished high school with a life list of better than three hundred species, and I was more interested in science than Manley was, and yet I never aimed my binoculars at anything but the night sky.

One reason I didn't was that the best birders at my high school were serious potheads and acid users. Also, most of them were boys. Birding wasn't necessarily nerdy (nerds didn't come to school tripping), but the scene associated with it was not my idea of galvanic. Of romantic. Tramping in woods and fields for ten hours, steadily looking at birds, not communicating about anything but birds, spending a Saturday that way, was strikingly akin, as a social experience, to getting baked.

Which itself may have been one reason why, in the year following my introduction to the veery, as I began to bird more often and stay out longer, I had a creeping sense of shame about what I was doing. Even as I was learning my gulls and sparrows, I took care, in New York, not to wear my binoculars on a strap but to carry them cupped discreetly in one hand, and if I brought a field guide to the park, I made sure to keep the front cover, which had the word *BIRDS* in large type, facing inward. On a trip to London, I mentioned to a friend there, a book editor who is a very stylish dresser, that I'd seen a green woodpecker eating ants in Hyde Park, and he made a horrible face and said, "Oh, Christ, don't tell me you're a twitcher." An American friend, the editor of a design magazine, also a sharp dresser, similarly clutched her head when I told her I'd been looking at birds. "No, no, no, no, no, no," she said. "You are *not* going to be a birdwatcher."

"Why not?"

"Because birdwatchers—*ucch*. They're all so—*ucch*."

"But if *I'm* doing it," I said, "and if I'm not that way—"

"But that's the thing!" she said. "You're going to *become* that way. And then I won't want to see you anymore."

She was talking in part about accessories, such as the elastic harness that birders attach to their binoculars to minimize neck strain and whose nickname, I'm afraid, is "the bra." But the really disturbing specter that my friend had in mind was the undefended sincerity of birders. The nakedness of their seeking. Their so-public twitching hunger. The problem was less acute in the shady Ramble (whose recesses, significantly, are popular for both daytime birding and nighttime gay cruising); but in highly public New York places, like on the Bow Bridge, I couldn't bear to hold my binoculars to my eyes for more than a few seconds. It was just too embarrassing to feel, or to imagine, that my private transports were being witnessed by better-defended New Yorkers.

And so it was in California that the affair really took off. My furtive hour-long get-togethers gave way to daylong escapes that I openly spent birding, wearing the bra. I set the alarm clock in the Californian's cabin for gruesomely early hours. To be juggling a stick shift and a thermos of coffee when the roads were still gray and empty, to be out ahead of everyone, to see no headlights on the Pacific Coast Highway, to be the only car pulled over at Rancho del Oso State Park, to already be on site when the birds were waking up, to hear their voices in the willow thickets and the salt marsh and the meadow whose scattered oaks were draped with epiphytes, to sense the birds' collective beauty imminent and findable in there: what a pure joy this all was. In New York, when I hadn't slept enough, my face ached all day; in California, after my first morning look at a foraging grosbeak or a diving scoter, I felt connected to a nicely calibrated drip of speed. Days passed like hours. I moved at the same pace as the sun in the sky; I could almost feel the earth turning. I took a short, hard nap in my car and woke up to see two golden eagles arrogantly working a hillside. I stopped at a feed lot to look for tricolored and yellow-headed blackbirds amid a thousand more plebeian birds, and what I saw instead, when the multitude wheeled into defensive flight, was a merlin coming to perch on a water tower. I walked for a mile in promising woods and saw basically nothing, a retreating thrush, some plain-Jane kinglets, and then, just as I was remembering what a monumental waste of time birding was, the woods came alive with songbirds, something fresh on every branch, and for the next fifteen minutes each birdlike movement in the woods was a gift to be unwrapped—western wood-pewee, MacGillivray's warbler, pygmy nuthatch—and then, just as suddenly, the wave was gone again, like inspiration or ecstasy, and the woods were quiet.

Always, in the past, I'd felt like a failure at the task of being satisfied by nature's beauty. Hiking in the West, my wife and I had sometimes found our way to summits unruined by other hikers, but even then, when the hike was perfect, I would wonder, "Now what?" And take a picture. Take another picture. Like a man with a photogenic girlfriend he didn't love. As if, unable to be satisfied myself, I at least might impress somebody else later on. And when the picture-taking finally came to feel just too pointless, I took mental pictures. I enlisted my wife to agree that such-and-such vista was incredible, I imagined myself in a movie with this vista in the background and various girls I'd known in high school and college watching the movie and being impressed with me; but nothing worked. The stimulations remained stubbornly theoretical, like sex on Prozac.

Only now, when nature had become the place where birds were, did I finally get what all the fuss was about. The California towhee that I watched at breakfast every morning, the plainest of medium-small brown birds, a modest ground dweller, a giver of cheerful, elementary chipping calls, brought me more pleasure than Half Dome at sunrise or the ocean shoreline at Big Sur. The California towhee generally, the whole species, reliably uniform in its plumage and habits, was like a friend whose energy and optimism had escaped the confines of a single body to animate roadsides and back yards across thousands of square miles. And there were 650 other species that bred in the United States and Canada, a population so varied in look and habitat and behavior—cranes, hummingbirds, eagles, shear-waters, snipe—that, taken as a whole, they were like a companion with an inexhaustibly rich personality. They made me happy like nothing outdoors ever had.

My response to this happiness, naturally, was to worry that I was in the grip of something diseased and bad and wrong. An addiction. Every morning, driving to an office I'd borrowed in Santa Cruz, I would wrestle with the urge to stop and bird for "a few minutes." Seeing a good bird made me want to stay out and see more good birds. Not seeing a good bird made me sour and desolate, for which the only cure was, likewise, to keep looking. If I did manage not to stop for "a few minutes," and if my work then didn't go well, I would sit and think about how high the sun was getting and how stupid I'd been to chain myself to my desk. Finally, toward noon, I would grab my binoculars, at which point the only way not to feel guilty about blowing off a workday was to focus utterly on the rendezvous, to open a field guide against the steering wheel and compare, for the twentieth time, the bill shapes and plumages of Pacific and red-throated loons. If I got stuck behind a slow car or made a wrong turn, I swore viciously and jerked the wheel and crushed the brakes and floored the accelerator.

I worried about my problem, but I couldn't stop. On business trips, I took whole personal days for birding, in Arizona and Minnesota and Florida, and it was here, on these solitary trips, that my affair with birds began to compound the very grief I was seeking refuge from. Phoebe Snetsinger, in her pointedly titled memoir, *Birding on Borrowed Time*, had described how many of the great avian haunts she'd visited in the eighties were diminished or destroyed by the late nineties. Driving on new arteries, seeing valley after valley sprawled over, habitat after habitat wiped out, I became increasingly distressed about the plight of wild birds. The ground dwellers were being killed by the tens of millions by domestic and feral cats, the low fliers were getting run down on ever-expanding exurban roads, the medium fliers were dismembering themselves on cellphone towers and wind turbines, the high fliers were colliding with brightly lit skyscrapers or mistaking rain-slick parking lots for lakes or landing in "refuges" where men in boots lined up to shoot them. On Arizona roads, the least fuel-efficient vehicles identified themselves with American flags and bumper-sticker messages like IF YOU CAN'T FEED 'EM, DON'T BREED 'EM. The Bush Administration claimed that Congress never intended the Endangered Species Act to interfere with commerce if local jobs were at stake—in effect, that endangered species should enjoy federal protection only on land that nobody had any conceivable commercial use for. The country as a whole had become so hostile to the have-nots that large numbers of the have-nots themselves now voted against their own economic interests.

The difficulty for birds, in a political climate like this, is that they are just profoundly poor. To put it as strongly as possible: they subsist on bugs. Also on worms, seeds, weeds, buds, rodents, minnows, pond greens, grubs, and garbage. A few lucky species—what birders call "trash birds"—cadge a living in urban neighborhoods, but to find more interesting species it's best to go to sketchy areas: sewage ponds, landfills, foul-smelling mudflats, railroad rights-of-way, abandoned buildings, tamarack swamps, thornbushes, tundra, weedy slashes, slime-covered rocks in shallow lagoons, open plains of harsh sawgrass, manure pits on dairy farms, ankle-turning desert washes. The species that reside in and around these bird ghettos are themselves fairly lucky. It's the birds with more expensive tastes, the terns and plovers that insist on beachfront housing, the murrelets and owls that nest in old-growth forests, that end up on endangered-species lists.

Birds not only want to use our valuable land, they're also hopelessly unable to pay for it. In Minnesota, north of Duluth, on an overcast morning when the temperature was hovering near ten, I saw a clan of white-winged crossbills, a flock of muted reds and golds and greens, crawling all over the apex of a snowy spruce tree. They weighed less than an ounce apiece, they'd been outdoors all winter, they were flashy in their feather coats, the spruce cones were apparently delicious to them, and even as I envied them their sociability in the snow I worried for their safety in the for-profit future now plotted by the conservatives in Washington. In this future, a small percentage of people will win the big prize—the Lincoln Navigator, the mansion with a two-story atrium and a five-acre lawn, the second home in Laguna Beach—and everybody else will be offered electronic simulacra of luxuries to wish for. The obvious difficulty for crossbills in this future is that crossbills don't *want* the Navigator. They don't *want* the atrium or the amenities of Laguna. What crossbills want is boreal forests where they can crack open seed cones with their parrot-of-the-northland bills. When our atmospheric carbon raises global temperatures by another five degrees, and our remaining unlogged boreal forests succumb to insects emboldened by the shorter winters, and cross-bills run out of places to live, the "ownership society" isn't going to help them. Their standard of living won't be improvable by global free trade. Not even the pathetic state lottery will be an option for them then.

In Florida, at the Estero Lagoon at Fort Myers Beach, where, according to my guidebook, I was likely to find "hundreds" of red knots and Wilson's plovers, I instead found a Jimmy Buffett song playing on the Holiday Inn beachfront sound system and a flock of gulls loitering on the white sand behind the hotel. It was happy hour. As I was scanning the flock, making sure that it consisted entirely of ring-billed gulls and laughing gulls, a tourist came over to take pictures. She kept moving closer, absorbed in her snapshots, and the flock amoebically distanced itself from her, some of the gulls hopping a little in their haste, the group murmuring uneasily and finally breaking into alarm cries as the woman bore down with her pocket digital camera. How, I wondered, could she not see that the gulls only wanted to be left alone? Then again, the gulls didn't seem to mind the Jimmy Buffett. The animal who most clearly wanted to be left alone was me. Farther down the beach, still looking for the promised throngs of red knots and Wilson's plovers, I came upon a particularly charmless stretch of muddy sand on which there were a handful of more common shorebirds, dunlins and semipalmated plovers and least sandpipers, in their brownish-gray winter plumage. Camped out amid high-rise condos and hotels, surveying the beach in postures of sleepy disgruntlement, with their heads scrunched down and their eyes half shut, they looked like a little band of misfits. Like a premonition of a future in which all birds will either collaborate with modernity or go off to die someplace quietly. What I felt for them went beyond love. I felt outright identification. The well-adjusted throngs of collaborator birds in South Florida, both the trash pigeons and trash grackles and the more stately but equally tame pelicans and cormorants, all struck me now as traitors. It was this motley band of modest peeps and plovers on the beach who reminded me of the human beings I loved best—the ones who didn't fit in. These birds may or may not have been capable of emotion, but the way they looked, beleaguered there, few in number, my outcast friends, was how I felt. I'd been told that it was bad to anthropomorphize, but I could no longer remember why. It was, in any case, anthropomorphic only to see yourself in other species, not to see them in yourself. To be hungry all the time, to be mad for sex, to not believe in global warming, to be shortsighted, to live without thought of your grandchildren, to spend half your life on personal grooming, to be perpetually on guard, to be compulsive, to be habit-bound, to be avid, to be unimpressed with humanity, to prefer your own kind: these were all ways of being like a bird. Later in the evening, in posh, necropolitan Naples, on a sidewalk outside a hotel whose elevator doors were decorated with huge blowups of cute children and the monosyllabic injunction SMILE, I spotted two disaffected teenagers, two little chicks, in full Goth plumage, and I wished that I could introduce them to the brownish-gray misfits on the beach.

A few weeks after I heard Al Gore speak at the Ethical Culture Society, I went back to Texas. According to my new AviSys 5.0 bird-listing software, the green kingfisher that I'd seen in the last hour of my trip with Manley had been my 370th North American bird. I was close to the satisfying milestone of four hundred species, and the easiest way to reach it without waiting around for spring migration was to go south again.

I also missed Texas. For a person with a bird problem, there was something oddly reassuring about the place. The lower Rio Grande Valley contained some of the ugliest land I'd ever seen: dead flat expanses of industrial farming and downmarket sprawl bisected by U.S. Route 83, which was a jerry-rigged viaduct flanked by three-lane frontage roads, Whataburgers, warehouses, billboards suggesting VAGINAL REJUNVENATION and FAITH PLEASES GOD and DON'T DUMP ("Take your trash to a landfill"), rotten town centers where only the Payless shoe stores seemed to be in business, and fake-adobe strip malls so pristinely bleak it was hard to tell if they were still being built or had already opened and gone bankrupt. And yet, to birds, the valley was a Michelin three-star destination: Worth a Journey! Texas was the home of President Bush and House Majority Leader Tom DeLay, neither of whom had ever been mistaken for a friend of the environment; its property owners were famously hostile to federal regulation; and yet it was the state where, with some serious driving, you could tally 230 species of bird in a single day. There were thriving Audubon Societies, the world's biggest birding-tour operator, special campgrounds and RV parks for birders, twenty annual birding festivals, and the Great Texas Coastal Birding Trail, which snaked for twenty-one hundred miles around petrochemical installations and supertanker hulls and giant citrus farms, from Port Arthur to Laredo. Texans didn't seem to lose much sleep over the division between nature and civilization. Even ardent bird lovers in Texas referred to birds collectively as "the resource." Texans liked to use the oxymoron "wildlife management." They were comfortable with hunting and viewed birding as basically a nonviolent version of it. They gave me blank, dumbfounded looks when I asked them if they identified with birds and felt a kinship with them, or whether, on the contrary, they saw birds as beings very different from themselves. They asked me to repeat the question.

I flew into McAllen. After revisiting the refuges I'd hit with Manley and bagging specialties like the pauraque (No. 374), the elf owl (No. 379), and the fulvous whistling-duck (No. 383), I drove north to a scrap of state land where the black-capped vireo (No. 388) and golden-cheeked warbler (No. 390), two endangered species, were helpfully singing out their locations. Much of my best birding, however, took place on private land. A friend of a friend's friend gave me a tour of his eight-thousand-acre ranch near Waco, letting me pick up three new inland sandpiper species on wetlands that the federal government had paid him to create. On the King Ranch, whose land holdings are larger than Rhode Island and include a hundred thousand acres of critical coastal habitat for migrating songbirds, I paid $119 for the opportunity to see my first ferruginous pygmy-owl and my first northern beardless-tyrannulet. North of Harlingen, I visited other friends of friends' friends, a pediatric dentist and his wife who had created a private wildlife refuge for themselves on five thousand acres of mesquite. The couple had dug a lake, converted old hunting blinds to nature-photography blinds, and planted big flower beds to attract birds and butterflies. They told me about their efforts to reeducate certain of their landowning neighbors who, like my father in the seventies, had been alienated by environmental bureaucrats. To be Texan was to take pride in the beauty and diversity of Texan wildlife, and the couple believed that the conservationist spirit in most Texan ranchers just needed a little coaxing out.

This, of course, was an axiom of movement conservatism—if you get government off people's back, they'll gladly take responsibility—and it seemed to me both wishful and potentially self-serving. At a distance, in New York, through the fog of contemporary politics, I probably would have identified the dentist and his wife, who were Bush supporters, as my enemies. But the picture was trickier in close-up. For one thing, I was liking all the Texans I met. I was also beginning to wonder whether, poor though birds are, they might prefer to take their chances in a radically privatized America where income distribution is ever more unequal, the estate tax is repealed, and land-proud Texan ranchers are able to preserve their oak mottes and vast mesquite thickets and lease them out to wealthy hunters. It certainly was pleasant to bird on a private ranch! Far away from the picnickers and the busloads of schoolkids! Far from the bikers, the off-roaders, the dog walkers, the smoochers, the dumpers, the partyers, the bird-indifferent masses! The fences that kept them out were no impediment to thrushes and wrens.

It was on federal property, though, that I got my four-hundredth species. In the village of Rockport, on Aransas Bay, I boarded a shallow-draft birding boat, the *Skimmer*, which was captained by an affable young outdoorsman named Tommy Moore. My fellow passengers were some eager older women and their silent husbands. If they'd been picnicking in a place where I had a rarity staked out, I might not have liked them, but they were on the *Skimmer* to look at birds. As we cut across the bay's shallow, cement-gray waters and bore down on the roosting site of a dozen great blue herons—birds so common I hardly noticed them anymore—the women began to wail with astonishment and pleasure: "Oh! Oh! What magnificent birds! Oh! Look at them! Oh my God!"

We pulled up alongside a very considerable green salt marsh. In the distance, hip-deep in salt grass, were two adult whooping cranes whose white breasts and long, sturdy necks and russet heads reflected sunlight that then passed through my binoculars and fell upon my retinas, allowing me to claim the crane as my No. 400. One of the animals was bending down as if concerned about something in the tall grass; the other seemed to be scanning the horizon anxiously. Their attitude reminded me of parent birds I'd seen in distress elsewhere—two bluejays in the Ramble fluttering in futile, crazed rage while a raccoon ate their eggs; a jittery, too-alert loon sitting shoulder-deep in water by the side of a badly flooded Minnesota lake, persisting in incubating eggs that weren't going to hatch—and Captain Moore explained that harm appeared to have befallen the yearling child of these two cranes; they'd been standing in the same place for more than a day, the young crane nowhere to be seen.

"Could it be dead?" one of the women asked.

"The parents wouldn't still be there if it had died," Moore said. He took out his radio and called in a report on the birds to the Aransas National Wildlife Refuge office, which told him that the chief crane biologist was on his way out to investigate.

"In fact," Moore told us, stowing the radio, "there he is."

Half a mile away, on the far side of a shallow salt pool, keeping his head low and moving very slowly, was a speck of a human figure. The sight of him there, in stringently protected federal territory, was disconcerting in the way of a boom mike dipping into a climactic movie scene, a stagehand wandering around behind Jason and Medea. Must humankind insert itself into *everything*? Having paid thirty-five dollars for my ticket, I'd expected a more perfect illusion of nature.

The biologist himself, inching toward the cranes, alone in his waders, didn't look as if he felt any embarrassment. It was simply his job to try to keep the whooping crane from going extinct. And this job, in one sense, was fairly hopeless. There were currently fewer than 350 wild whooping cranes on the planet, and although the figure was definitely an improvement on the 1941 population of 22, the long-term outlook for any species with such a small gene pool was dismal. The entire Aransas reserve was one melted Greenland ice cap away from being suitable for waterskiing, one severe storm away from being a killing field. Nevertheless, as Captain Moore cheerfully informed us, scientists had been taking eggs from the cranes' nests in western Canada and incubating them in Florida, where there was now a wholly manufactured second flock of more than thirty birds, and since whooping cranes don't naturally know the way to migrate (each new generation learns the route by following its parents), scientists had been trying to teach the cranes in Florida to follow an airplane to a second summering site in Wisconsin . . .

To know that something is doomed and to cheerfully try to save it anyway: it was a characteristic of my mother. I had finally started to love her near the end of her life, when she was undergoing a year of chemotherapy and radiation and living by herself. I'd admired her bravery for that. I'd admired her will to recuperate and her extraordinary tolerance of pain. I'd felt proud when her sister remarked to me, "Your mother looks better two days after abdominal surgery than I do at a dinner party." I'd admired her skill and ruthlessness at the bridge table, where she wore the same determined frown when she had everything under control as when she knew she was going down. The last decade of her life, which started with my father's dementia and ended with her colon cancer, was a rotten hand that she played like a winner. Even toward the end, though, I couldn't tolerate being with her for more than three days at a time. Although she was my last living link to a web of Midwestern relations and traditions that I would begin to miss the moment she was gone, and although the last time I saw her in her house, in April 1999, her cancer was back and she was rapidly losing weight, I still took care to arrive in St. Louis on a Friday afternoon and leave on a Monday night. She, for her part, was accustomed to my leavings and didn't complain too much. But she still felt about me what she'd always felt, which was what I wouldn't really feel about her until after she was gone. "I hate it when Daylight Savings Time starts while you're here," she told me while we were driving to the airport, "because it means I have an hour less with you."

As the *Skimmer* moved up the channel, we were able to approach other cranes close enough to hear them crunching on blue crabs, the staple of their winter diet. We saw a pair doing the prancing, graceful, semiairborne dance that gets them sexually excited. Following the lead of my fellow passengers, I took out my camera and dutifully snapped some pictures. But all of a sudden—it might have been my having reached the empty plateau of four hundred species—I felt weary of birds and birding. For the moment at least, I was ready to be home in New York again, home among my kind. Every happy day with the Californian made the dimensions of our future losses a little more grievous, every good hour sharpened my sadness at how fast our lives were going, how rapidly death was coming out to meet us, but I still couldn't wait to see her: to set down my bags inside the door, to go and find her in her study, where she would probably be chipping away at her interminable e-mail queue, and to hear her say, as she always said when I came home, "So? What did you see?"

Notes

Introduction

1. E. B. White, "Mr. Forbush's Friends," *New Yorker,* February 26, 1966, 42–66.

2. Edward Howe Forbush, *Birds of Massachusetts and Other New England States,* 1:342, 158, 342, 315.

3. Forbush, 338–41.

3. Birding in Traffic

1. "Poetry," first published in *Others: A Magazine of the New Verse* 5, no. 6, (July 1919), 5. Reprinted in her collection *Observations*, edited and with an Introduction by Linda Leavell (New York: Farrar, Straus and Giroux, 1924), 27.

4. Buried Birds

1. Sea birds like petrels, shearwaters, and prions are drawn to lights much as moths are. Ships are a beacon to them, and they can become stunned or trapped on deck. In breeding areas, ships are required to shutter their windows at night to prevent such incidents.

2. Richard Nelson, *The Island Within*, Vintage, 1991, 153.

3. Karkalla is the local name of the ice plant, or *Carpobrotus rossii.*

4. The *Protector III*, a sealing vessel, beached on New Island in 1969. The thin-billed prion (*Pachyptila belcheri*) nests on New Island, digging ~3m burrows into the hillsides.

Nocturnal nesters, the adults leave the chicks in the burrows during the day while they forage at sea. New Island is most likely the most important breeding spot for these birds, representing 30% of the world population.

5. Rabbits, rats, goats—these are often introduced creatures, not native to the islands on which pelagic birds breed. In North America, Antarctica, and elsewhere, they are pests and harmful to the birds. What a shock to be in a place where their presence is benign.

5. The Problem with Pretty Birds

1. Wallace Stevens, "Of Mere Being," in *The Palm at the End of the Mind: Selected Poems and a Play*, ed. Holly Stevens (New York: Knopf, 1967).

2. Roger Tory Peterson, *A Field Guide to the Birds*, 4th ed. (New York: Houghton Mifflin, 1980), 274.

3. Annie Dillard, *Pilgrim at Tinker Creek* (New York: Harper & Row, 1974), 105.

4. David Foster Wallace, *The Pale King* (New York: Little, Brown, 2011), 374.

5. David Allen Sibley, *The Sibley Field Guide to Birds of Eastern North America* (New York: Knopf, 2003), 364.

6. John Keats, "Ode to a Nightingale," originally published in 1819. From Poetry Foundation, https://www.poetryfoundation.org/poems/44479/ode-to-a-nightingale.

7. Rainer Maria Rilke, *Duino Elegies*, 4, 1923.

8. Scott Russell Sanders, *Staying Put: Making a Home in a Restless World* (Boston: Beacon Press, 1993), xv.

7. Crane, Water, Change

1. Brina Kessel, "Migration of Sandhill Cranes, *Grus canadensis*, in East-Central Alaska, with Routes through Alaska and Western Canada," *The Canadian-Field Naturalist*, July–Sept 1984. This article remains the seminal paper on sandhill crane migration thirty years after its original publication.

2. The Cornell Lab of Ornithology: https://www.allaboutbirds.org/guide/sandhill_crane/lifehistory.

3. Kim Heacox, *The Only Kayak* (Guilford, CT: Lyons Press, 2005).

4. From Richard Nelson's radio show *Encounters,* an episode on cranes.

5. NOAA releases an annual Arctic Report Card with data on Arctic climate trends, https://arctic.noaa.gov/Report-Card; this essay cites the 2014 report.

6. Michelle Nijhuis, "A Storm Gathers for North American Birds," *Audubon Magazine*, September–October 2014, online version at https://www.audubon.org/magazine/september-october-2014/a-storm-gathers-north-american-birds. Audubon's Climate Report includes a wealth of information, research and projections on birds, and excellent essays synthesizing the data: Audubon Birds & Climate Change Report, https://www.audubon.org/menu/birds-climate-change-report?page=1.

7. Melanie Smith, telephone conversation with the author, November 3, 2014.

8. Of special concern are wintering grounds in Platte River country, where up to 80% of North America's migratory sandhills stop.

9. Melanie Smith, telephone conversation with the author, November 3, 2014.

10. William Stafford, "Watching Sandhill Cranes," in *Even In Quiet Places* (Lewiston, ID: Confluence Press, 1997), 67. Copyright 1996, the Estate of William Stafford. Reprinted with the permission of The Permissions Company, LLC on behalf of Confluence Press, www.confluencepress.com.

11. Sharman Apt Russell, "Meet the Beetles," *Orion Magazine*, October 22, 2014, 32.

8. The Black and White

1. Richard Rhodes, *John James Audubon: The Making of an American* (New York: Alfred A. Knopf, 2004), 11.

2. Rhodes, *Audubon*, 12.

9. One Single Hummingbird

1. Thomas Moore, *Care of the Soul: A Guide for Cultivating Depth and Sacredness in Everyday Life* (New York: Harper Perennial, 2016), 112.

2. J. A. Baker, *The Peregrine* (New York: New York Review of Books Classics, 2004), 14.

3. Baker, *The Peregrine*, 41.

4. Baker, *The Peregrine*, 121.

5. Henry David Thoreau, "Walking," originally published in *Atlantic Monthly*, June 1862.

6. Sallie McFague, *Super, Natural Christians: How We Should Love Nature* (Minneapolis: Fortress Press, 2000), 155.

11. Wild Swans

1. John Bull and John Farrand Jr., *The Audubon Society Field Guide to Birds* (New York: Alfred A. Knopf, 1977), 464.

2. Rachel Carson and Linda J. Lear, *Lost Woods: The Discovered Writing of Rachel Carson* (Boston: Beacon Press, 1998), 96.

3. James MacKillop, *A Dictionary of Celtic Mythology* (New York: Oxford University Press, 2000), 394.

4. W. B. Yeats, "The Wild Swans at Coole," *The Collected Poems* (New York: Macmillan Publishing Company, 1974), 129.

12. The Snowy Winter

1. Letter from Melville to his editor Evert Duyckinck in 1850, quoted in Howard Payton Vincent, *The Trying-Out of Moby-Dick* (Ohio: Kent State University Press, 1980), 42.

2. Team eBird, "Arctic Wanderers—Snowy Owl Invasion 2013," eBird website, December 11, 2013, https://ebird.org/news/gotsnowies2013.

3. John Schwartz, "A Bird Flies South, and It's News," *New York Times*, January 31, 2014, https://www.nytimes.com/2014/02/01/us/influx-of-snowy-owls-thrills-and-baffles-birders.html.

15. In the Eyes of the Condor

1. John Nielsen, *Condor: To the Brink and Back—The Life and Times of One Giant Bird* (New York: HarperCollins, 2006), 48.

16. Little Brown Birds

1. R. Rubin, "Not Far from Forsaken," *New York Times Magazine*, April 9, 2006, http://www.nytimes.com/2006/04/09/magazine/not-far-from-forsaken.html.

2. Deborah Sontag and Robert Gebeloff, "The Downside of the Boom," *New York Times*, November 22, 2014, http://www.nytimes.com/interactive/2014/11/23/us/north-dakota-oil-boom-downside.html.

3. Carl T. Montgomery and Michael B. Smith, "Hydraulic Fracturing: History of an Enduring Technology," *Journal of Petroleum Technology* 12 (2010): 26–32, https://doi.org/10.2118/1210-0026-JPT.

4. Montgomery and Smith, "Hydraulic Fracturing."

5. Hans M. Kristenson and Matt Korda, "Worldwide Deployments of Nuclear Weapons, 2017," *Bulletin of the Atomic Scientists* 73.5 (2017): 289–297.

6. Eric Schlosser, *Command and Control: Nuclear Weapons, the Damascus Accident, and the Illusion of Safety* (New York: Penguin, 2013).

17. The Keepers of the Ghost Bird

1. William Strachey, "A true repertory of the wracke, and redemption of Sir Thomas Gates Knight; upon, and from the llands of the Bermudas . . ., " in *Purchas his Pilgrimes*, compiled by Samuel Purchase, vol. 4, chap. 6 (London, 1625). British Library, bl.uk/collection-items/stracheys-a-true-reportory-of-the-wreck-in-bermuda.

2. "No Longer Extinct. The Cahow: Saved from Hog, Rat, and Man." *New York Times*, December 2, 1973, nytimes.com/1973/12/02/archives/no-longer-extinct-the-cahow-saved-from-hog-rat-and-man-cahow-david.

3. *Rare Bird* (documentary film), directed by Lucinda Spurling, written by Jack McDonald and Lucinda Spurling. Afflare Films, March 18, 2006, Bermuda.

4. *Rare Bird*, directed by Lucinda Spurling.

5. *Rare Bird*, directed by Lucinda Spurling.

6. Elizabeth Gerhman, "Unraveling the Mysteries," in *Rare Birds: The Extraordinary Tale of the Bermuda Petrel and the Man Who Brought It Back from Extinction* (Boston: Beacon Press, 2012).

7. Bobbii Cartwright and Liz Nash, *History and Rationale of the Living Museum Project; Guide to Nonsuch Island "Living Museum" Nature Reserve*, in consultation with David Wingate. Pamphlet, Bermuda Zoological Society, Bermuda, April 2001.

8. Lewis Thomas, "The Lives of a Cell," in *The Lives of A Cell: Notes of a Biology Watcher* (New York: Penguin Books, 1978).

18. The Hour (or Two) before the Dawn

1. And he did it! You can read all about his ride to raise awareness of the perils of global climate change: Kroodsma, David. 2014. *The Bicycle Diaries. My 21,000-mile Ride for the Climate*. RFC Press, San Francisco, California.

2. It took a few months to tidy up loose ends at the university, but on December 31, 2003, as the clock struck midnight, I was officially *FREE*! Many years later now, I can still feel the exhilaration of that second.

3. F. M. Chapman, *Handbook of Birds of Eastern North America*, 2nd ed. (New York: Dover Publications,1939).

20. Guardian of the Garden

1. Karl Ove Knausgaard, *My Struggle: Book 2: A Man in Love*, trans. Don Bartlett. Translation edition June 3, 2014 (New York: Farrar, Straus and Giroux), 590.

2. Terry Tempest Williams, *When Women Were Birds: Fifty-four Variations on Voice* (New York: Picador, 2012), 186. Complete quotation: "I am not Louis's mother, but I have become a mother, which is an unspoken agreement to be forever vulnerable."

23. Death and the Rose-breasted Grosbeak

1. Edward O. Wilson, *Biophilia* (Cambridge, MA: Harvard University Press, 1984), 6.
2. Gretel Ehrlich, *The Solace of Open Spaces* (New York: Viking Penguin, 1985), 63.
3. Iris Murdoch quoted in Jeremy Mynott, *Birdscapes: Birds in Our Imagination and Experience*, (Princeton: Princeton University Press, 2009), 298.
4. Lewis, *A Grief Observed* (Harper & Row, 1961), 3.
5. Edna St. Vincent Millay, *Collected Poems*, ed. Norma Millay (New York: Harper & Row, 1956), 448.
6. Joseph Severn, quoted in Walter Jackson Bate, *John Keats*, (Cambridge, MA: Harvard University Press, 1964), 255.
7. John Keats, quoted in Bate, *John Keats*, 425.
8. John Keats, "Ode to a Nightingale," in *Complete Poems*, ed. Jack Stillinger (Cambridge, MA: Harvard University Press, 1978).
9. John Keats, quoted in Bate, *John Keats*, 506.
10. David Bottoms, "Under the Vulture-Tree," in *Birds in the Hand: Fiction and Poetry About Birds*, ed. Kent Nelson and Dylan Nelson (North Point Press, 2004), 47.
11. Keats, "Ode to a Nightingale," *Complete Poems*, 369.
12. Mary Oliver, "The Return," *What Do We Know: Poems and Prose Poems* (Da Capo Press, 2003), 8.
13. Keats, "What the Thrush Said," *Complete Poems*, 235.
14. Keats, "Ode on Melancholy," *Complete Poems*, 374.

25. Nest Watcher

1. Kathleen Dean Moore, *Great Tide Rising: Towards Clarity and Moral Courage in a Time of Planetary Change* (Berkeley, CA: Counterpoint Press, 2016), 37.

Contributor Biographies

Christina Baal is the artist behind *Drawing 10,000 Birds* and can be met showing her art at bird festivals across the country. Since graduating from Bard College in 2014, she has designed artwork for major bird festivals and events, including Space Coast Birding and Wildlife Festival, the World Series of Birding, and the Biggest Week in American Birding. Her work has appeared in nature- and bird-themed exhibitions, including *Birdland and the Anthropocene* (Baltimore, Maryland, 2017) and *Drawn to Nature* (Audubon, Pennsylvania, 2018, 2019). An artist-in-residence at Puembo Birding Garden, Ecuador, and two-time artist-in-residence at the Ucross Foundation, in Wyoming, she also had a solo show at the Ucross Foundation Art Gallery in 2017 entitled *The Universal Language of Birds*. Baal incorporates art into her work as an environmental educator to fuse the arts and sciences in order to inspire others to love and protect the natural world. Her most recent project is writing and illustrating a book on mythological birds with the support of a William Mullen Grant from Bard College. www.drawingtenthousandbirds.com.

Thomas Bancroft is a writer and photographer in love with wild places. His vivid photography and sound recordings weave through his stories about his adventures in wilderness and inspire passion for its protection. After earning his PhD in ornithology, he dedicated his career to the preservation of wild spaces throughout the United States and Latin America, and now works in countries around the world. He led a team at the interface of science and policy for the Wilderness Society and National Audubon, and now serves on boards of multiple organizations. He has been published in scientific journals and popular magazines, and his photographs displayed in the Smithsonian and Burke Museums. In his spare time he can be found filling the bird feeders surrounding his light-filled home in Seattle, Washington, teaching birding and nature, and hiking the wildest old-growth forests he can find.

K. Bannerman lives on Vancouver Island, Canada, where she writes short stories, novels, and screenplays. Her novels include the cosmic-horror-romance *Love and Lovecraft*, the werewolf tale *The Tattooed Wolf*, and the historical murder mystery *Bucket of Blood*. She also hosts the weekly history podcast *Northwest by Night*.

R. A. Behrstock, formerly a fisheries biologist teaching at Humboldt State University in Northern California, moved to Texas to become a cofounder of Ben Feltner's Peregrine Tours. After six years with Peregrine, he spent twelve years as a senior leader for WINGS Birding Tours, concentrating on the southern U.S. and Latin America. With Fermata, Inc., Behrstock evaluated sites for seven Texas birding trails, eventually coauthoring *Birds of Houston, Galveston, and the Upper Texas Coast*, and the Upper Texas Coast volume of *Finding Birds on the Great Texas Coastal Birding Trail*, as well as an introductory field guide to southwestern dragonflies and damselflies. He has authored or coauthored about sixty popular and scientific papers concerning fishes, birds, dragonflies, grasshoppers, and butterflies in the U.S. and Latin America and more than 1,200 of his photos have been used in books, magazines, museum exhibits, calendars, and digital field guides. Behrstock lives in southeastern Arizona with his partner, Karen. He has just retired from leading tours in the U.S. and Latin America for Naturalist Journeys. Currently, he serves as a director of Pollinator Corridors, a nonprofit in southeastern Arizona, and is documenting insects pollinating milkweeds.

Richard Bohannon is a cartographer, sociologist, and amateur birder who teaches in the College of Individualized Studies at Metropolitan State University in Saint Paul, Minnesota. He is the editor of *Religions and Environments: A Reader in Religion, Nature, and Ecology* and co-editor of *Grounding Religion: A Field Guide to Religion and Ecology*. Bohannon recently finished a project quantifying the effects of habitat fragmentation on breeding bird populations in North Dakota's Bakken region (results in the *Annals of the American Association of Geographers)*, out of which also emerged his contribution to this book.

Elizabeth Bradfield is a writer/naturalist and author of the poetry collections *Once Removed*, *Approaching Ice*, and *Interpretive Work*, and *Toward Antarctica,* which combines her photographs with brief, hybrid essays. Her poems and essays have appeared in the *New Yorker, West Branch, Orion,* and many anthologies. Founder and editor in chief of Broadsided Press, she lives on Cape Cod, works as a naturalist locally as well as on expedition ships, and teaches creative writing at Brandeis University.

In her essay "Buried Birds," Bradfield interweaves a more traditional essayistic form with Japanese *haibun. Haibun* recounts scenes from everyday life or travels in diary-like prose scattered with small poems, usually haiku. The best examples are found in the writings of Bashō, the seventeenth-century master of the form, whose *Back Roads to Far Towns* recounts the journey he took with his disciple Sora into remote, northern Japan. As with much classical Japanese poetry and art, Bashō's work is infused with allusions to poets, stories, figures, and cultural tropes that add resonance to his writing and which his contemporaries would have recognized. Given our diverse contemporary world, Bradfield has included notes with her essay to offer the modern reader a rendering of that undercurrent.

Christine Byl is the author of *Dirt Work: An Education in the Woods*, a book about trail crews, tools, wildness, gender, and labor. It was shortlisted for the 2014 Willa Literary Award in nonfiction. Her fiction and essays have appeared in *Glimmer Train, The Sun, Crazyhorse,* and *Brevity*, among other journals and anthologies. Byl has made her living as a professional trail builder for the past twenty-four years working on public lands from the sub-Arctic to Patagonia. She lives in Interior Alaska with her family and spends as much time as possible exploring open spaces by foot, bike, skis, boat, and dogs.

Susan Cerulean is a writer, naturalist, and activist. Her book *Coming to Pass: Florida's Coastal Islands in a Gulf of Change* was awarded a Gold Medal for Florida Nonfiction in 2016. Her first nature memoir, *Tracking Desire: A Journey after Swallow-tailed Kites*, was named Editors' Choice by *Audubon* magazine. In summer 2020, her new book, *I Have Been Assigned the Single Bird: A Daughter's Memoir,* will be released by the University of Georgia Press. Cerulean edited *Between Two Rivers: Stories from the Red Hills to the Gulf*, *Unspoiled: Writers Speak for Florida's Coast*, and *The Book of the Everglades*. With her husband, she divides her time between Tallahassee and Indian Pass, Florida. She is president of the Friends of St. Vincent National Wildlife Refuge. Her website is www.comingtopass.com.

Sara Crosby grew up in St. Louis, Missouri, and studied at the University of Iowa before getting an MFA at The New School in New York. She has taught and been an administrator at the City University of New York for sixteen years, but would always rather be watching birds, writing essays, or walking her dog, Andy. Her writing has been published in *The Believer*, *Fourth Genre*, *Ninth Letter*, *PEN America*, *Pembroke Magazine*, and others. "The Black and White" earned a Notable Essays mention in *The Best American Essays*, 2010.

Jenn Dean holds an MFA in Literature and Nonfiction from the Bennington Writing Seminars. A portion of her unfinished memoir, *House of My Sleepless Nights*, was published in *Salamander,* and her interview with the writer Jane Brox appeared in the *Writer's Chronicle*. Awarded a Millay Colony residency, she's been a featured artist in SoundFalls, an evening of music and stories, in Carnation, Washington, and was a finalist for the Lamar York Nonfiction Prize. A former member of the Birdnote.org production team, she has served as a trip host on BirdNote's tour of the Galapagos. She is currently working on a book about the Snoqualmie Valley of Washington, where she lives, called *Letters from the Valley of the Moon*.

Dean's essay "The Keepers of the Ghost Bird" was originally published in a longer form through *Massachusetts Review*'s Working Titles series as an e-book. The essay was a 2016 finalist in the *New Millennium* Writing Awards and also won the 2018 John Burroughs Nature Essay Award. Her website is www.jenndean.com.

Rachel Dickinson is a writer and artist who lives in perpetually overcast central New York. Her work has appeared in numerous print and online publications, including *The Atlantic, Audubon, National Geographic Traveler, Salon, Catapult,* and *Smithsonian.* She's also the author of several nonfiction books, including *Falconer on the Edge: A Man, His Birds, and the Vanishing Landscape of the American West* and *The Notorious Reno Gang: The Wild Story of the West's First Brotherhood of Thieves, Assassins, and Train Robbers.*

Katie Fallon is the author of the nonfiction books *Vulture: The Private Life of an Unloved Bird* and *Cerulean Blues: A Personal Search for a Vanishing Songbird*, which was a finalist for the Reed Writing Award for outstanding writing on the southern environment.

Fallon is also the author of two books for children. Her essays and articles have appeared in a variety of journals and magazines, including *Fourth Genre*, *River Teeth*, *Ecotone*, and *Bark Magazine.* She has taught creative writing at Virginia Tech and West Virginia University and teaches in the low-residency MFA programs at Chatham University and West Virginia Wesleyan College. She is a founder of the nonprofit Avian Conservation Center of Appalachia and has served as president of the Mountaineer Chapter of the National Audubon Society. Her first word was "bird."

Jonathan Franzen is a novelist, essayist, journalist, translator, and screenwriter. He is a member of the American Academy of Arts and Letters and the author of five novels, including *The Corrections, Freedom,* and *Purity,* and five volumes of nonfiction, most recently *The End of the End of the Earth.* An ardent bird-watcher, he has served on the board of the American Bird Conservancy since 2008, and he has received the EuroNatur Award for his advocacy on behalf of European birds. He lives in Santa Cruz, California.

Andrew Furman is a professor of English at Florida Atlantic University and teaches in its MFA program in creative writing. His writing frequently engages with the Florida outdoors and has appeared in such publications as *Oxford American*, the *Southern Review*, *Ecotone*, the *Wall Street Journal*, *Poets & Writers*, the *Chronicle of Higher Education*, *Agni Online*, *Terrain.org*, *Flyway*, and the *Florida Review*. He is the author, most recently, of the novel *Goldens Are Here* and the memoir *Bitten: My Unexpected Love Affair with Florida*, which was named a finalist for the ASLE Environmental Book Award. A new novel, *Jewfish*, is forthcoming in 2020.

Tim Gallagher is an award-winning author and magazine editor. He was editor in chief of *Living Bird* magazine at the Cornell Lab of Ornithology for twenty-six years. In 2004 Gallagher reported seeing an Ivory-billed Woodpecker in Arkansas and wrote a best-selling book about the experience, *The Grail Bird: Hot on the Trail of the Ivory-billed Woodpecker*. He is also the author of *Falcon Fever*, a memoir about his lifelong fascination with birds of prey; *Imperial Dreams: Tracking the Imperial Woodpecker through the Wild Sierra Madre*; *Parts Unknown: A Naturalist's Journey in Search of Birds and Wild Places*; *Wild Bird Photography*; and *Born to Fish*.

David Gessner is the author of eleven books, including *All the Wild That Remains: Edward Abbey, Wallace Stegner, and the American West*; *Return of the Osprey*, which follows the return of these raptors to Cape Cod; *Sick of Nature*; the award-winning *The Tarball Chronicles*, and *Leave It as It Is: A Journey through Theodore Roosevelt's American Wilderness*. Gessner serves as chair of the Creative Writing Department at the University of North Carolina, Wilmington, where he founded the literary journal *Ecotone*.

Renata Golden began birding as a young girl on the South Side of Chicago. When her mother gave names to Northern Cardinals, Blue Jays, and American Robins, Golden was hooked on the idea of naming the things she saw in the natural world. She realized she needed to expand her vision on her first trip out West, when a friend spotted a Western Tanager in a ponderosa pine. She now knows a Western Kingbird from a Cassin's Kingbird and is convinced she can tell a Common Nighthawk from a Lesser Nighthawk midflight. Golden holds an MFA from the Creative

Writing Program at the University of Houston. She has been published in the literary journals *Creative Nonfiction: True Stories*, *Chautauqua*, *About Place Journal*, *Terrain.org*, and *Muse/A Journal*, as well as several newspapers and magazines. Her essay in *Chautauqua* about facing her fear in a wild cave was nominated for a 2020 Pushcart Prize. "Guardian of the Garden" is part of a longer book project titled *Thin Places: The Nature of the Chiricahua Mountains*.

Ursula Murray Husted spends most of her time drawing pictures, making comics, daydreaming about boats, and feeding the cats. She has traveled to the Arctic to see Snow Buntings and polar bears. She is the author of the graphic novel *A Cat Story*, about cats, art, and the meaning of life. Her website is www.ursulamurrayhusted.com.

Eli J. Knapp loves to be where the wild things are. He cherishes kayaks, tree branches, Land Rover roofs, his back deck, anywhere that offers a promising vantage point. His observations and adventures often morph into narratives, published in outlets like *New York State Conservationist*, *Pennsylvania Magazine*, *Birdwatcher's Digest*, and most recently a book, *The Delightful Horror of Family Birding: Sharing Nature with the Next Generation*. When not bushwhacking through thickets and writing about it, Knapp teaches courses in the Intercultural Studies and Biology departments at Houghton College, in western New York. His research spans the conservation gamut, emerging from a three-year stint living in Serengeti National Park, in Tanzania, where he studied the coexistence of people and wildlife around protected areas. Each spring he takes students to East Africa where he directs a study abroad semester. Knapp best enjoys experiential education opportunities, seeking special places for his students to become active participants in learning. He lives with his wife, Linda, and three children, Ezra, Indigo, and Willow, in Fillmore, New York. Nestled under pines that whistle in the wind, his three children make certain he remains where the wild things are.

Donald Kroodsma is a world-renowned authority on birdsong and professor emeritus of ornithology at the University of Massachusetts, Amherst. As a research scientist, he published widely on birdsong for more than fifty years, receiving lifetime achievement awards from the American Ornithologists' Union, the Wilson Ornithological Society, and the American

Birding Association. More recently he has authored books that introduce the general public to birdsong: *The Singing Life of Birds*, which won the John Burroughs Medal; *The Backyard Birdsong Guides*; *Birdsong by the Seasons*; *Listening to a Continent Sing: Birdsong by Bicycle from the Atlantic to the Pacific*; and *Birdsong for the Curious Naturalist*. He lives in Hatfield, Massachusetts.

J. Drew Lanham is a native of Edgefield and Aiken, South Carolina. In his twenty years at Clemson University, where he holds an endowed chair as an Alumni Distinguished Professor and is also an Alumni Master Teacher, he has worked to understand how forest management impacts wildlife and how human beings think about nature. As a Black American Lanham is intrigued with how culture and ethnic prisms can bend perceptions of nature and its care. His "Connecting the Conservation Dots" and "Coloring the Conservation Conversation" messages have been delivered internationally as calls for increased focus on inclusion, diversity, and passion in the environmental/conservation movement. Lanham is a widely published author and award-nominated poet, writing about his experiences as a birder, hunter, and wild, wandering soul. His book *The Home Place: Memoirs of a Colored Man's Love Affair with Nature* and his collection of poetry *Sparrow Envy* speak to his passion for nature. Lanham was named the third Poet Laureate of Edgefield County in 2018.

John R. Nelson is the author of *Flight Calls: Exploring Massachusetts through Birds*. He has contributed essays and stories about birds and nature to the *Antioch Review,* the *Gettysburg Review,* the *Harvard Review*, *Harvard Magazine,* the *Missouri Review,* the *New England Review,* and birding journals in the U.S. and Great Britain. His essay on birds and dance, "Brolga the Dancing Crane Girl," was awarded *Shenandoah*'s Thomas Carter Prize for Nonfiction in 2012, and his essay "Funny Bird Sex," published in the *Antioch Review*, was awarded a 2018 Pushcart Prize. Nelson chairs the conservation and education committee of the Brookline Bird Club and founded the Association of Massachusetts Bird Clubs. A professor emeritus at North Shore Community College, he lives in Gloucester, Massachusetts.

Rob Nixon is the prize-winning author of four books, including *Dreambirds: The Natural History of a Fantasy* and *Slow Violence and the Environmentalism of the Poor*. He holds the Barron Family Professorship in Humanities and the Environment at Princeton University. Nixon has received many awards, including a MacArthur Foundation Peace and Security Fellowship.

Jonathan Rosen is the author of two novels, *Eve's Apple* and *Joy Comes in the Morning*, and two nonfiction books, *The Talmud and the Internet: A Journey between Worlds* and *The Life of the Skies: Birding at the End of Nature*. His essays and articles have appeared in the *New York Times*, the *New Yorker*, *The Atlantic*, the *Wall Street Journal,* and numerous anthologies. He created the culture section of *The Forward* newspaper, which he oversaw for ten years, and created and edited the Jewish Encounters Book Series, published by Schocken, an imprint of Random House, and Nextbook Inc. He lives and birds in New York City and is writing a book for Penguin Press about friendship and madness.

Alison Townsend is the author of *The Persistence of Rivers: An Essay on Moving Water*, which won the Jeanne Leiby Prose Chapbook Award, and two books of poetry, *The Blue Dress*, selected for the Marie Alexander Prose Poem Series at White Pine Press, and *Persephone in America*, Crab Orchard Open Poetry Competition winner, as well as two poetry chapbooks. Her poetry and nonfiction appear in journals such as *Bellingham Review*, *Chautauqua*, *Chattahoochee Review, Crab Orchard Review*, *Southern Review*, and *Water~Stone*. Her work has appeared in *Best American Poetry* and has received a Pushcart Prize, five *Best American Essays* "Notable" mentions, a Wisconsin Literary Arts Grant, and the University of Wisconsin-Whitewater Chancellor's Regional Literary Award, among other honors. Emerita professor of English at the University of Wisconsin-Whitewater, she lives with her climate activist husband in the farm country outside Madison, the inspiration for her forthcoming essay collection, *American Lonely: A Natural History of My Search for Home*.

Alison Világ was six when her parents took her to look at ducks. Little did they know that day would lead to a life with the birds, both professionally and as a pastime. From the Great Lakes region, where she grew up, Világ has traveled out—from Alaska's Pribilof Islands to Borneo—in search of birds. Világ received her BA in environmental writing from Unity College in Maine. Through guiding, writing, and photography, she strives to connect people with nature. Világ works at Whitefish Point Bird Observatory in Michigan's Upper Peninsula.

Credits

Elizabeth Bradfield, "The Blinds Must be Closed by Dusk" and "New Island, Falklands," which are part of "Buried Birds," originally appeared in Elizabeth Bradfield, *Toward Antarctica*, copyright Red Hen Press, 2019. Reprinted with the permission of Red Hen Press. "Buried Birds" was first published in *Tahoma Literary Review*, issue 17, 2020.

Christine Byl, "Crane, Water, Change: A Migratory Essay," was commissioned and published online by Denali National Park and appeared in an earlier form in the Denali Climate Anthology at https://www.nps.gov/dena/getinvolved/dca_byl.htm. Reprinted with the author's permission.

Susan Cerulean, "Nest Watcher," has been previously published in Susan Cerulean, *I Have Been Assigned the Single Bird*. Copyright 2020, University of Georgia Press. Reprinted with permission of University of Georgia Press.

Sara Crosby, "The Black and White," first appeared in *Fourth Genre: Explorations in Nonfiction*, volume 11, number 2, Michigan State University. Copyright 2009, Sara Crosby. Reprinted with the author's permission.

Jenn Dean, "The Keepers of the Ghost Bird," was originally published in a longer form through *Massachusetts Review*'s Working Titles series as an e-book. Copyright 2016 by Jenn Dean. Reprinted with the author's permission.

Katie Fallon, "Nighthawks: Lake Perez" first appeared in the Shaver's Creek Environmental Center's online "The Creek Journals" as part of the Shaver's Creek Long-term

Ecological Reflections Project (https://www.shaverscreek.org/about-us/long-term-ecological-reflections-project/). Reprinted with permission of Shaver's Creek Environmental Center.

Jonathan Franzen, "My Bird Problem," was originally published, in a somewhat different form, in the *New Yorker*, August 9, 2005. Reprinted with the author's permission.

Andrew Furman, "The Problem with Pretty Birds," first appeared in *Terrain.org* on December 23, 2015. Reprinted with the author's permission.

Donald Kroodsma, "The Hour (or Two) before the Dawn," is adapted from chapter 24 of *Listening to a Continent Sing: Birdsong by Bicycle from the Atlantic to the Pacific*. Princeton University Press. Copyright 2016, Donald Kroodsma. Reprinted by permission of Princeton University Press.

John R. Nelson, "Death and the Rose-breasted Grosbeak," was originally published in issue 41 of *Harvard Review* and adapted for *Flight Calls*, University of Massachusetts Press, 2019. Reprinted by permission of University of Massachusetts Press.

Jonathan Rosen, "Birding in Traffic," originally appeared as "Lemon Zest" in the City Section of the *New York Times* on February 4, 2008. Reprinted with the author's permission.

Alison Townsend, "Wild Swans," first appeared in *Chautauqua*, issue 11, "Wonders of the World." The essay also received a "Notable" mention in *Best American Essays, 2015*. Reprinted with the author's permission.